Business Models and Reliable Operation of Virtual Power Plants

Heping Jia · Xuanyuan Wang · Xian Zhang ·
Dunnan Liu

Business Models and Reliable Operation of Virtual Power Plants

Springer

Heping Jia
North China Electric Power University
Beijing, China

Xuanyuan Wang
State Grid Jibei Electric Power Co., Ltd.
Beijing, China

Xian Zhang
Beijing Power Exchange Center Co., Ltd.
Beijing, China

Dunnan Liu
North China Electric Power University
Beijing, China

ISBN 978-981-19-7848-7 ISBN 978-981-19-7846-3 (eBook)
https://doi.org/10.1007/978-981-19-7846-3

This Springer imprint is published by the registered company Springer Nature Singapore Pte Ltd.
The registered company address is: 152 Beach Road, #21-01/04 Gateway East, Singapore 189721, Singapore

Preface

The world faces unavoidable multiple climate hazards over the next two decades with global warming of 1.5 °C. Greenhouse gas emissions from human activities are the most significant driver of observed climate change where emissions come from the energy sector accounts for almost three-quarters. Therefore, the solutions to make decarbonization are using electricity from renewable energy. In 2021, global renewable energy installed capacity rises 9%. The total share of global renewable generation capacity in electricity production rose from 36.6% in 2020 to 38.3% in 2021. However, due to the intermittent, volatility, and uncertainties of renewable energy generations, there is an urgent need to improve the power system flexibility. With the development of distributed energy resources, information and communication technologies, flexible resources in demand side can be used to deal with the volatility of new energies. As a solution to solve the problem of clean energy consumption and green energy transformation, virtual power plant (VPP) technology is of great significance to global energy transformation to realize the goal of low carbon by aggregating distributed energy resources.

The main content of this book summarizes the research results of Dr. Heping Jia, Associate Professor of North China Electric Power University, Dr. Xuanyuan Wang, Director General of Technology and Innovation Division in State Grid Jibei Electric Power Co., Ltd., Dr. Xian Zhang, Director of New Energy Exchange Department in Beijing Power Exchange Center Co., Ltd., Dr. Dunnan Liu, Professor of North China Electric Power University, and their collaborators in this field. This book systematically studies the business models of virtual power plants with different practices and develops the reliable operation models of virtual power plants under different operating scenarios, which has important and practical guiding significance for the operation mechanism and optimal operation of virtual power plants.

The structure of the book is as follows. Chapter 1 summarizes the background knowledge of climate change and the emerging of virtual power plants by aggregating distributed energy resources. Chapter 2 introduces the short-term load forecasting for distributed energy resources as the basic analysis for virtual power plants. In Chap. 3, clustering forecasting of virtual power plants aggregated with electric vehicles is presented considering meteorological factors. In Chap. 4, several business models and practices of virtual power plants participating in peak shaving energy market, frequency control market, and comprehensive energy services are provided. In Chap. 5, an integrated electricity/heat demand response modeling approach for virtual power plants with comfort of consumers is presented. In Chap. 6, the bidding strategy and dispatching scheme of virtual power plants participating in auxiliary service market coordinating with energy storage allocation is given. In Chap. 7, a dynamic time-sharing pricing model for virtual power plants is provided based on deep deterministic policy gradient reinforcement learning algorithm. Chapter 8 introduces a reliable operation model for power systems with virtual power plants utilizing reliability network equivalent and time-sequential simulation techniques. In Chap. 9, an operating reliability evaluation framework for multi-state power systems with virtual power plants and wind power is presented considering malfunctions of cyber-systems. In the book, Chaps. 1 and 4 are jointly written by Dr. Heping Jia and Dr. Xuanyuan Wang; Chaps. 2, 3, 5, 6 and7 are jointly written by Dr. Heping Jia, Dr. Xian Zhang, and Prof. Dunnan Liu; Chaps. 8 and 9 are written by Dr. Heping Jia.

This book provides an insight of virtual power plants by aggregating distributed energy resources in smart grids. We believe this book will offer the related researchers a better understanding on business models and reliable operation of virtual power plants.

Beijing, China

Heping Jia
Xuanyuan Wang
Xian Zhang
Dunnan Liu

Acknowledgements

This book summarizes our research on business models and reliable operation of virtual power plants from 2017 to 2022. Many people contributed to this book. We offer our tonnes of thanks to Prof. Yonghua Song in University of Macau, Prof. Yi Ding from Zhejiang University, and Prof. Yanbin Li from North China Electric Power University. We also thank Mr. Weiye Wang, Ms. Lingxiang Wang, Ms. Qian Li, Ms. Min Yan, Ms. Yue Zhang, and Mr. Jian Wei for their careful examination of the entire book. Finally, we appreciate the assistance and help of the staff at Springer. This book is supported in part by National Natural Science Foundation (Grant No. 72001078), in part by the Young Elite Scientists Sponsorship Program by China Association for Science and Technology (Grant No. 2021QNRC001). We do appreciate their supports.

About This Book

This book systematically studies the business models of virtual power plants with different practices and develops the reliable operation models of virtual power plants under different operating scenarios, which has important and practical guiding significance for the operation mechanism and optimal operation of virtual power plants. The business models and reliable operation strategies proposed in this book are applicable to virtual power plants with different distributed energy resources. The business operations of virtual power plants participating in demand response and auxiliary service market are provided based on the load forecasting for distributed energy resources. The dynamic pricing strategy for virtual power plants is also presented. Moreover, the reliable operation of power systems integrated with virtual power plants is analyzed. The readers of this book are researchers, engineers, technicians, and graduate students engaged in the fields of energy Internet, electrical engineering, business administration, etc.

Contents

Chapter 1
Climate Change and Virtual Power Plants

For nearly a hundred years, many observations have shown that the Earth's climate is undergoing a significant change with global warming as the main feature. It is reported by the Intergovernmental Panel on Climate Change (IPCC) that global warming in the past 50 years was mainly caused by the warming effect of greenhouse gases such as CO_2, CH_4, N_2O from human activities [1]. In order to jointly address the challenge of climate change and slow down global warming, nearly 200 parties jointly adopted the Paris Agreement in December 2015, making action arrangements on how to tackle climate change after 2020. The main goal of this agreement is to keep global temperature increases in the twenty-first century within 2 °C higher than pre-industrial levels, and to seek to stay further within 1.5 °C [2]. Including the Paris Agreement, China is an active participant and practitioner of global climate change activities such as the United Nations Framework Convention on Climate Change (UNFCCC) and the Kyoto Protocol. In September 2020, China announced that it would aim to achieve peak CO_2 emissions before 2030 and carbon neutrality before 2060. The goals for peak CO_2 emissions and carbon neutrality constitute a strategic choice for the sake of humanity's future.

Global carbon emissions have grown rapidly from 2010 to 2020 as demonstrated in Fig. 1, reaching a record high of 34.04 billion tons in 2019 [3]. The huge negative impact of COVID-19 on global social production and life, global carbon dioxide emissions fell by 5.8% in 2020, close to 2Gt of carbon dioxide, the biggest drop since 1945 [3]. The impact of human activities is crucial. For example, burning fossil energy increases the release of CO_2 into the atmosphere, while deforestation weakens the carbon sink process, thus damaging the balance, leading to high CO_2 concentrations in the atmosphere and rising temperatures.

Globally, the energy sector including electricity, heat and transport produces 73.2% of its greenhouse gas emissions [4] as illustrated in Fig. 2. Therefore, driving net zero emissions in the energy sector is the key to tackling climate change. Specifically, for global CO_2 emissions as illustrated in Fig. 3, energy emissions mainly came from energy power generation and heating, transportation, manufacturing and construction, accounting for 43%, 26% and 17% in 2020, respectively [5].

H. Jia et al., *Business Models and Reliable Operation of Virtual Power Plants*,
https://doi.org/10.1007/978-981-19-7846-3_1

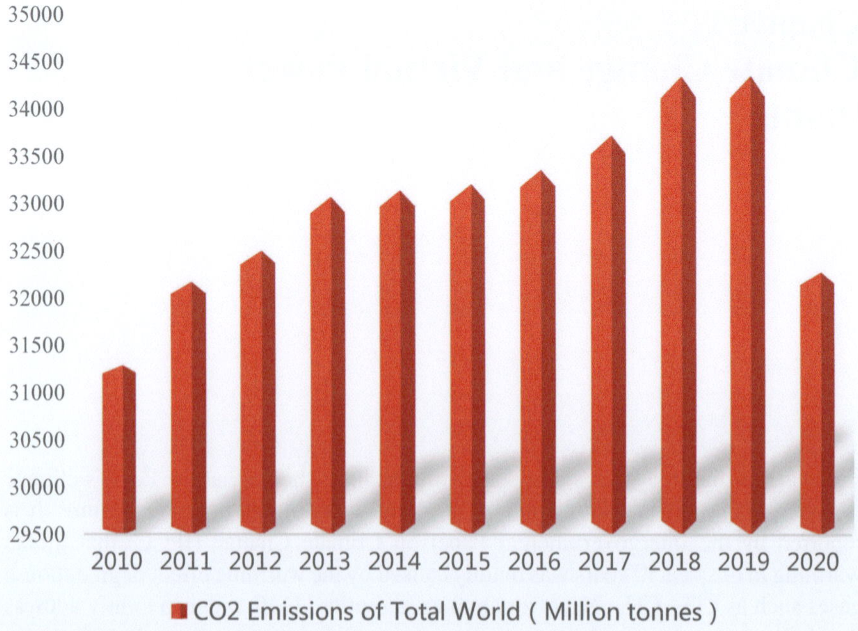

Fig. 1 CO_2 emissions of total world, 2010–2020 (https://www.bp.com/)

Fossil power generation is one of the largest sources of energy-related carbon emissions, accounting for 36% of total energy-related emissions, with renewable energy contributing the most to the decarbonization of the net zero power sector [6]. Note that the CO_2 emissions from power generation and heating were nearly half of the total emissions. Hence, reducing the supply of fossil energy for power generation and heating and increasing the power generation from renewable energy are the keys to dealing with climate change.

For the past decade, renewables have met a higher share of global electricity demand than in the previous year as illustrated in Fig. 4. In 2020, Non-hydro renewables generated an estimated 29.0% of global electricity up from 27.3% in 2019. The share of electricity generated by variable renewable electricity (wind power and solar PV) continued to rise in several countries around the world, which contributed more than 9% of global electricity in 2020. Moreover, much higher shares of variable renewable electricity were globally presented including Denmark with a share of 63%, Uruguay with a share of 43%, Ireland with a share of 38%, Germany with a share of 33%, and the United Kingdom with a share of 28% [7].

In the meanwhile, a record amount of new capacity of the installation of solar PV and wind power has been achieved as presented in Figs. 5 and 6. The world total capacity of solar PV achieved a record-breaking 139 GW of new installations, bringing total capacity to nearly 760 GW. The world total capacity of wind power achieved a record-breaking 93 GW of new installations, bringing total capacity onshore and offshore to nearly 743 GW. Driven by pending policy changes at the

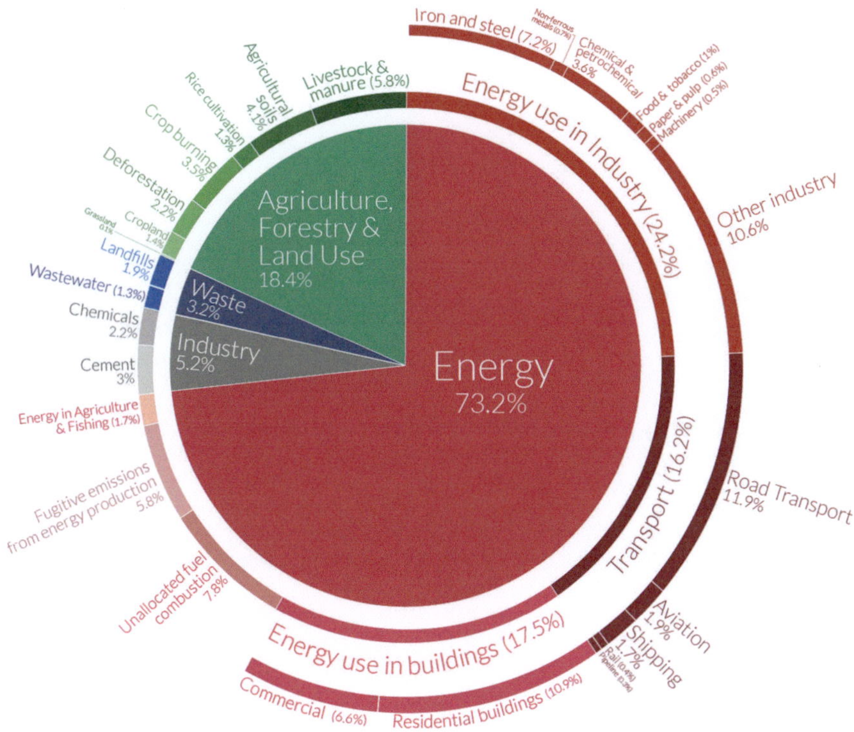

Fig. 2 The global greenhouse gas emissions by sectors (https://ourworldindata.org/emissions-by-sector)

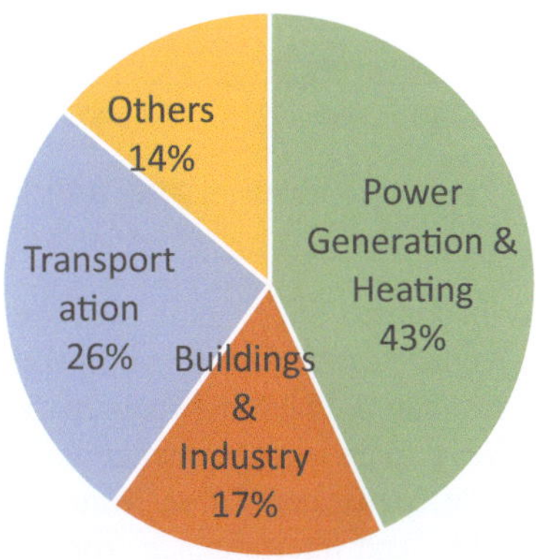

Fig. 3 CO_2 emissions sources by industries, 2020 (https://www.iea.org/)

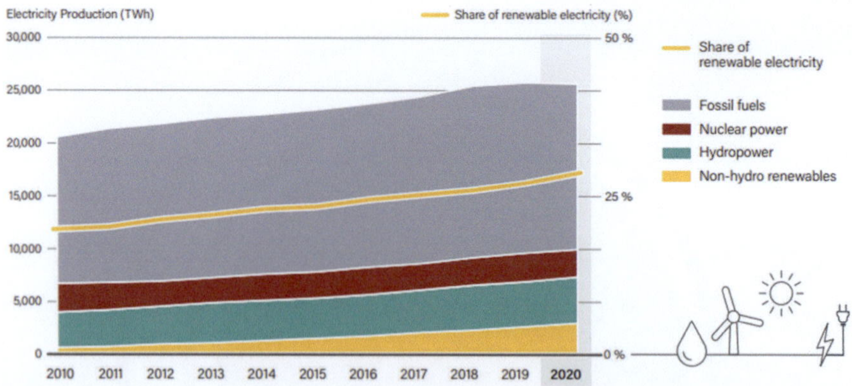

Fig. 4 Global electricity production by source, and share of renewables, 2010–2020 (https://www.ren21.net/)

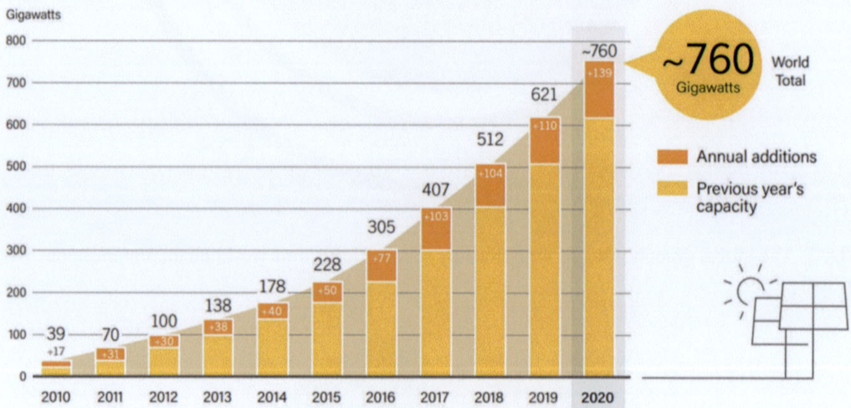

Fig. 5 Solar PV global capacity and annual additions, 2010–2020 (https://www.ren21.net/)

end of 2020 in China and the United States, both countries led the growth in solar PV and wind power with record years. Viewed over a slightly longer period, wind and solar capacity was more than doubled between 2015 and 2020, increasing by around 800 GW, which equated to an average annual increase of 18%.

With the increasing of electricity generation from wind power and solar PV, the characteristics of volatility, and intermittence from renewable energy generation are increasing the demand for flexible resources of power systems [8]. Flexible resources were defined in [9] as generation resources whose operations can be directly controlled (are dispatchable) and quickly start up, shut down, and ramp power output up and down.

Besides flexible resources from electricity generation side, with the development of distributed energy resources (DERs) including distributed generations (DGs),

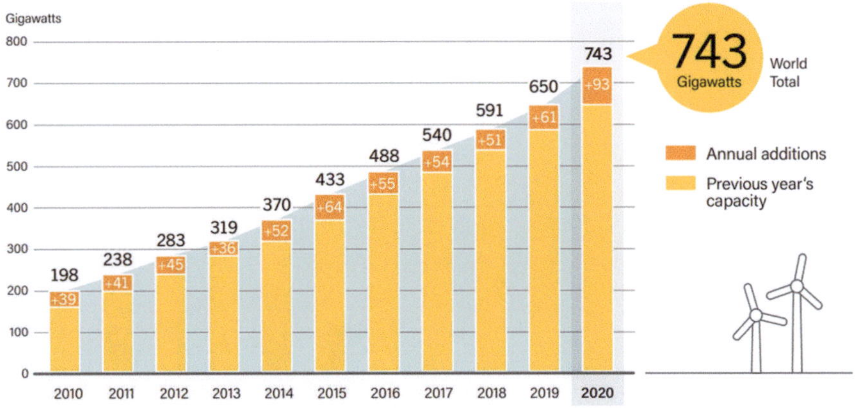

Fig. 6 Wind power global capacity and annual additions, 2010–2020 (https://www.ren21.net/)

controllable loads (CLs), electrical vehicles (EVs), electrical energy storages (EESs), etc., and information and communication technologies (ICTs), more distribution networks are changing from one-way power flow to bidirectional power flow [10]. Therefore, flexible resources can be practically extended to demand-side resources besides generation resources. Taking electrical vehicles (EVs) as an example, the overall global electric car stock reached 10 million in 2020 as illustrated in Fig. 7, with battery electric vehicles (BEVs) accounting for two-thirds of the world's electric car fleet. The theoretical charging/discharging power of BEVs was about 150 million kW, which was about 10% of the total installation capacity of global wind power and solar PV. Generally speaking, EVs incorporated into the power grid can not only charge but also discharge, thus providing upward and downward flexibility for the power grid. By optimizing the charging and discharging strategies, the widespread use of EVs is utilization of a kind of high-quality flexible resources. Due to the different charging and discharging characteristics of different types of EV batteries, appropriate charging and discharging strategies are implemented to achieve the balance of power supply and demand with significant traffic emission reduction effect.

There is an urgent need to improve the power system flexibility using flexible resources in demand side to deal with the volatility of new energies. However, the interaction potential of the practical flexible resources in demand side is limited. The interaction potential of a large scale of flexible resources is supposed to be explored. How to awaken a large number of demand-side flexible resources to solve the low-carbon-security-economy triangle contradiction in energy transformation?

As a solution to solve the problem of clean energy consumption and green energy transformation, virtual power plant (VPP) technology is of great significance to global energy transformation to realize the goal of low carbon. A VPP is a group of distributed energy resources which combine to function as a dispatchable unit [12]. More specifically, from the IEC standard (IEC TS 63189-2), VPP is defined as a

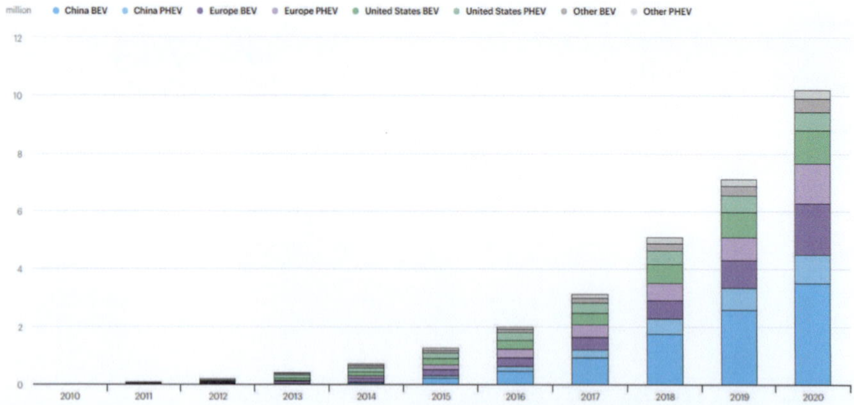

Fig. 7 Global electric car stock by region and mode, 2010–2020 (https://www.iea.org/)

party or system that realizes aggregation, optimization and control of decentralized generations, energy storage devices and controllable loads, which are not necessarily within the same geographical area, and facilitate the activities in power system operations and electricity market. For example, EVs are integrated in VPPs to provide flexibilities to power systems [13, 14].

Different from traditional thermal power plants, VPPs can reduce emissions through adjusting VPPs' power outputs avoiding power generated from fossil energy. Moreover, by aggregating various DERs, VPPs can not only participate in the electricity market bidding to sell electric energy as an electric energy provider, but also purchase electricity from the market as a load to meet its own electricity demand [15, 16].

References

1. Intergovernmental Panel on Climate Change, Climate change widespread, rapid, and intensifying, August 9, 2021, https://www.ipcc.ch/2021/08/09/ar6-wg1-20210809-pr/
2. https://www.un.org/en/climatechange/paris-agreement
3. BP energy, Statistical Review of World Energy 2021, https://www.bp.com/content/dam/bp/business-sites/en/global/corporate/pdfs/energy-economics/statistical-review/bp-stats-review-2021-full-report.pdf
4. Our World in Data, Our World in Data CO_2 and Greenhouse Gas Emissions, https://ourworldindata.org/emissions-by-sector
5. IEA, Global Energy Review 2021, https://www.iea.org/reports/global-energy-review-2021
6. IEA, Net Zero by 2050, https://www.iea.org/reports/net-zero-by-2050
7. REN21, Renewables 2021 Global Status Report, https://www.ren21.net/gsr-2021/
8. H. Jia, Y. Ding, Y. Song, C. Singh, M. Li, Operating reliability evaluation of restructured power systems considering flexible reserve provider on demand side. IEEE Trans. Smart Grid **10**(3), 3452–3464 (2019)
9. Briefing Paper: A Review of Current Issues with Long-Term Resource Adequacy, California Public Utilities Commission Energy Division and Policy and Planning Division (2013)

10. H. Hui, Y. Ding, Q. Shi, F. Li, Y. Song, J. Yan, 5G network-based Internet of Things for demand response in smart grid: a survey on application potential. Appl. Energy **257**, 1113972 (2020)
11. IEA, Electric Vehicle Sales Review 2021, https://www.iea.org/reports/electric-vehicles
12. https://electropedia.org/iev/iev.nsf/display?openform&ievref=617-04-27
13. X. Yang, Y. Zhang, A comprehensive review on electric vehicles integrated in virtual power plants. Sustain. Energy Technol. Assess. **48**, 101678 (2021)
14. J. Wang, C. Guo, C. Yu, Y. Liang, Virtual power plant containing electric vehicles scheduling strategies based on deep reinforcement learning. Electr. Power Syst. Res. **205**, 107714 (2022)
15. X. Liu, Research on bidding strategy of virtual power plant considering carbon-electricity integrated market mechanism. Int. J. Electr. Power Energy Syst. **137**, 107891 (2022)
16. A.R. Jordehi, A stochastic model for participation of virtual power plants in futures markets, pool markets and contracts with withdrawal penalty. J. Energy Storage **50**, 104334 (2022)

Chapter 2
Short-term Load Forecasting for DERs Based on CNN-LSTM with Attention Mechanism

The accurate load forecasting of DERs which are aggregated in VPPs is the fundamental idea for the reliable, economic and efficient operation of VPPs. Traditional forecasting methods cannot meet the requirements of high-precision load forecasting, while machine learning algorithm is widely utilized as one of the accurate schemes for short-term load forecasting. This chapter proposes an algorithm combined convolutional neural network (CNN) and long short-term memory (LSTM) with attention mechanism to forecast short-term load of DERs which are aggregated in VPPs. The proposed algorithm can reduce the loss of historical information and realize short-term load forecasting in different seasons. Moreover, the proposed load forecasting method is compared with other commonly used forecasting models to verify the forecasting accuracy of the proposed algorithm.

1 Introduction

To ensure the reliable operation of power systems, it is essential to have accurate short-term load forecasting. Moreover, the accurate load forecasting of DERs which are aggregated in VPPs is the fundamental idea for the reliable, economic and efficient operation of VPPs [1, 2]. The variations of short-term load are affected by various factors, such as weather. Traditional forecasting methods such as time series are not suitable for complex forecasting. Machine learning with convolutional neural network (CNN) as a kind of main forecasting models is the typical method for short-term load forecasting, and has been applied to many industries with high accuracy [3, 4]. The current methods for load forecasting mainly include support vector machine (SVM) and neural network [5], which can better deal with nonlinear problems. However, SVM methods have difficulties in dealing with large-scale data, while CNN methods can adequately solve the nonlinear problems that exist in large-scale load data. Therefore, CNN methods are widely used in the field for load forecasting [6]. In [7, 8], CNN models were utilized to extract features of the input data for

H. Jia et al., *Business Models and Reliable Operation of Virtual Power Plants*,
https://doi.org/10.1007/978-981-19-7846-3_2

load forecasting, noting that CNN is more accurate in learning nonlinear sequences of load data. However, when the load data is volatile and unstable, it is difficult for a single CNN model to learn the load variation pattern well. Li et al. [9] proposed a short-term load forecasting method based on long short-term memory (LSTM) network and illustrated the effectiveness of this method in extracting temporal characteristics of load values. Kong et al. [10] and Alhussein et al. [11] proposed a deep learning framework based on a combination of a CNN and LSTM. The proposed hybrid CNN-LSTM model uses CNN layers for feature extraction from the input data with LSTM layers for sequence learning. In [12], an adaptation of the load dispatching algorithm was proposed by combining the empirical mode decomposition, cascade-forward neural network and linear model through a fusion strategy for load forecasting.

However, algorithms, such as LSTM and RNN tend to lose sequence information when the input data sequence is too long, thus affecting the accuracy of load forecasting to a certain extent. To solve the problem, this chapter analyzes the load characteristics of DERs that aggregated in VPPs and proposes a combined model with both CNN and LSTM algorithms based on attention mechanism. In the proposed model, the LSTM implicit layer is given with different probability weights so as to enhance the influence of important information, which largely solves the problem of losing sequence information for a long sequence input and improves the accuracy of short-term load forecasting. Moreover, the proposed algorithm is compared with other typical methods through the analysis of a case study, which verifies the higher forecasting accuracy of the proposed algorithm.

2 Load Characteristics of DERs Aggregated in VPPs

The load of DERs aggregated in VPPs changes from moment to moment and is characterized by periodicity, fluctuation and randomness in time [13]. Periodicity is based on quarters, months, even hours and minutes, with regular changes, where small cycles may exist within large cycles. The fluctuation refers to the oscillation of load and the randomness refers to the fluctuation of load with a certain irregularity. The fluctuation magnitude is also different due to various factors.

As an example, the load curve of DERs aggregated in a VPP is illustrated in Fig. 1. The annual load curve of DERs has two peak loads in summer and winter, where the summer peak is significantly higher than that of the other seasons. Considering the great load variation in different seasons, this chapter considers the load characteristics of different seasons for short-term load forecasting.

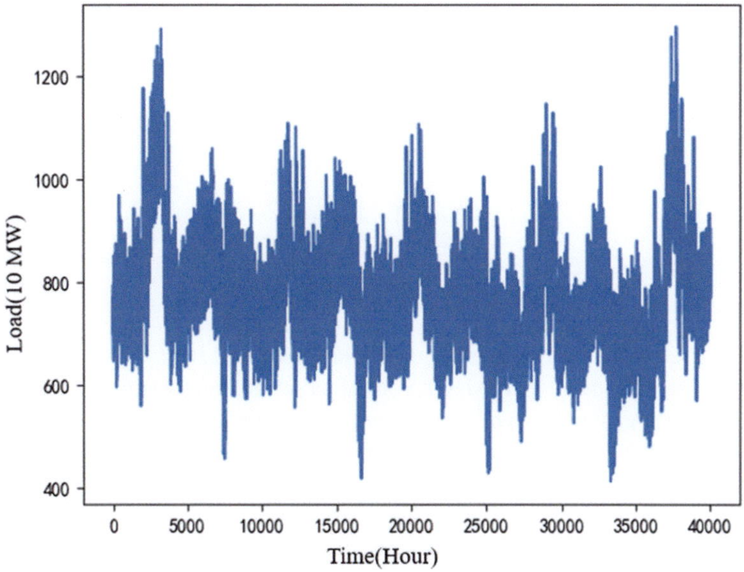

Fig. 1 Load curve of DERs aggregated in a VPP

3 Short-Term Load Forecasting Utilizing CNN-LSTM with Attention Mechanism

3.1 Brief Introduction of LSTM Algorithm

The design of LSTM is arguable from the logic gates of a computer. The structure of the LSTM includes forget gate, input gate and output gate. The forget gate selects the memory information of the previous moments and the new input information. Filtering by sigmoid function, the value of 1 means all information are through, while the value of 0 means all information cannot pass. Due to the presence of the forget gate, it is difficult for the LSTM to undergo gradient disappearance. The details of each LSTM calculation are formulated as follows:

The forget gate can be described as shown in (1).

$$f_t = \text{sigmoid}(w_f[h_t, x_t] + b_f) \tag{1}$$

where f_t represents the amount of information passed.

The input gate is formulated in (2)–(3).

$$i_t = \text{sigmoid}(w_i[h_{t-1}, x_t] + b_i) \tag{2}$$

$$\tilde{C}_t = \tanh(w_c[h_{t-1}, x_t] + b_c) \tag{3}$$

The cell state (information transfer) is shown in (4).

$$C_t = C_{t-1} * f_t + i_t * C_t \tag{4}$$

The output gate is presented in (5)–(6).

$$O_t = \text{sigmoid}(w_o[h_{t-1}, x_t] + b_o) \tag{5}$$

$$h_t = O_t * \tanh(C_t) \tag{6}$$

In (1)–(6), w_f, w_i, w_c, w_o are weighted indicators, and b_f, b_i, b_c, b_o are offsets.

Compared with the RNN model, LSTM adds several gate settings, especially the forget gate setting, which can filter the input information of the previous time period, thus retaining the key information and forget part of the unimportant information, which is the key to overcome the gradient disappearance. However, LSTM still has the phenomenon of losing important data information when the input sequence is too long. Therefore, CNN is utilized to process the original data and filter out part of the unimportant information, to improve the accuracy of load forecasting.

3.2 Brief Introduction of CNN Algorithm

The main role of CNN [14] is to perform feature extraction on data. CNN consists of convolutional layer, pooling layer, and fully connected layer. The convolution layer is the key to extract features, and the convolution kernel is responsible for extracting the corresponding data features. The more the number of convolution kernels, the more abstract the extracted features [16]. Pooling is divided into maximum pooling, average pooling, etc. In this chapter, the maximum pooling approach is utilized where the *RELU* activation function is embedded to ignore some unimportant features. The *RELU* activation function is a segmented linear function with the specific expression in (7). The role of the fully connected layer is to expand the pooled neurons into a one-dimensional vector form, thus making it easier to process the data. However, CNN has difficulty in learning the relationship between temporal data in forecasting. Therefore, enhancements of the typical methods of the RNN family are needed to combine CNN methods with LSTM methods.

$$\text{RELU} = \begin{cases} x & x \geq 0 \\ 0 & x < 0 \end{cases} \tag{7}$$

3.3 Attention Mechanism

The attention mechanism is a resource allocation mechanism mimicking the attention of the human brain, where the human brain focuses its attention on the area needing to be focused at a particular moment, reducing or even ignoring attention to other areas in order to obtain more detailed information. The attention mechanism suppresses other useless information, ignores irrelevant information and amplifies the desired information [16]. The attention mechanism increases the accuracy of the model by assigning sufficient attention to key information and highlighting the impact of important information by means of probabilistic assignment. The attention mechanism can effectively improve the situation that LSTM loses information due to too long sequences, while replacing the original method of randomly assigning weights by assigning probabilities.

3.4 The Procedures for Load Forecasting Based on CNN-LSTM with Attention Mechanism

The procedures for load forecasting based on CNN-LSTM with attention mechanism are illustrated in the flow chart in Fig. 2 as follows.

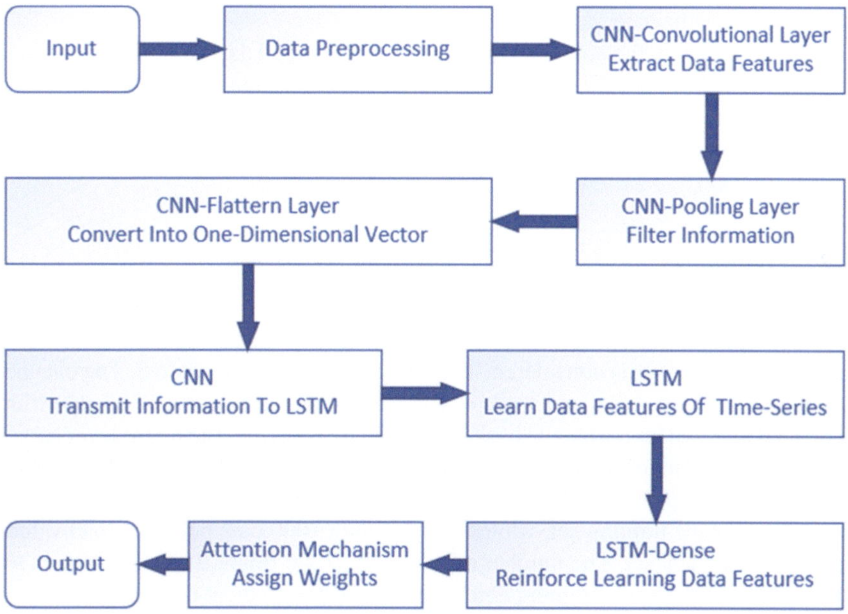

Fig. 2 The procedures for load forecasting based on CNN-LSTM with attention mechanism

(1) Data pre-processing: including normalization, removal of missing values, etc.
(2) Model training: the data from the training set is fed into the model for training, where features are extracted by the CNN layer; the LSTM learns the temporal features of the load values, while the Dropout method is used to suppress overfitting.
(3) Forecasting results output: the attention mechanism determines the output weight value to get the load prediction output value.

4 Case Study

4.1 The Time Series Analysis of Load Characteristics

Here, electricity load data with a one hour interval for a time period is analyzed. The history data is pre-processed including missing value filling and data normalization. Through the preliminary analysis of the load data, it can be found that with the change of seasons, the electricity load data have obvious cyclical patterns, where the load values in summer and winter are significantly higher than those in spring and autumn with larger fluctuations relative to the load values in spring and autumn. Therefore, this chapter adopts different seasons for electric load forecasting. The time series of different seasons for electricity load are illustrated in Fig. 3 where the horizontal coordinates are the data points corresponding to different seasons, and the vertical coordinates are the load values.

In order to verify the smoothness of the electricity load data in different seasons, this chapter utilizes augmented Dickey-Fuller (ADF) test and *P*-value test, as shown in Table 1. The data for four seasons passes the ADF test and the p-value test, which is easier to extract load features and filter irrelevant information using CNN.

4.2 Parameter Setting of the Proposed Algorithm

In this chapter, the proposed CNN-LSTM algorithm with one convolutional layer and one pooling layer is considered. The convolution kernel is set to 5. The channel width is set to be (6, 1, 1), and stride = 1 in the pooling layer. The LSTM algorithm is divided into two layers and its back-connected dense layer. The dense layer serves to enhance the role of learning data features, with 16 neurons in each layer and 1 neuron in the output layer. The first 9000 data points of load values in each season are utilized as the training set, while the remaining 1000 data points of load values are used as the test set. The number of training is 1000 times. The batch size is set to be 8000 and the adam training algorithm is used for gradient descent.

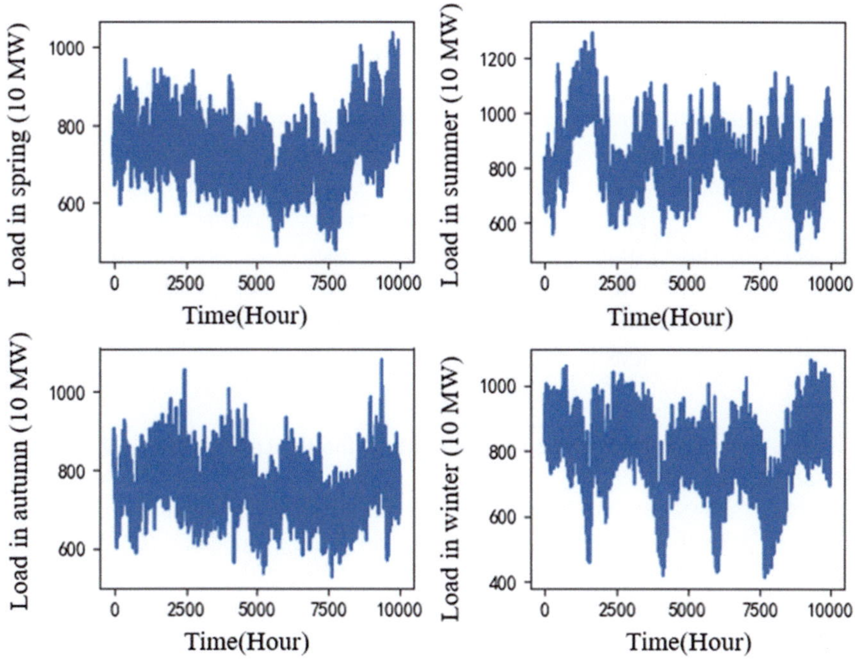

Fig. 3 The characteristic of electricity loads in different seasons

Table 1 ADF test and P value test for electricity loads

Season	ADF inspection	1%	5%	10%	P-value
Spring	−8.2158	−3.4310	−2.8618	−2.5669	0.0000
Summer	−4.6118	−3.4310	−2.8618	−2.5669	0.0001
Autumn	−6.8758	−3.4310	−2.8618	−2.5669	0.0000
Winter	−4.8758	−3.4310	−2.8618	−2.5669	0.0001

4.3 Comparison of Load Forecasting Errors

In this chapter, the mean absolute percentage error (MAPE) and root mean square error (RMSE) are adopted to present the load forecasting errors of different methods, as shown in (8)–(9).

$$\mathrm{MAPE} = \frac{1}{N}\sum_{i=1}^{N}\frac{\left|\hat{y}_i - y_i\right|}{y_i} \tag{8}$$

$$\mathrm{RMSE} = \sqrt{\frac{1}{N}\sum_{i=1}^{N}(y_i - \hat{y}_i)^2} \tag{9}$$

Table 2 Error comparison of different load forecasting methods

Date (test set)	Method	CNN	LSTM	CNN-LSTM	SVM	Random forest	Proposed method
2017/3/29–2017/5/10	MAPE	2.562%	2.996%	3.014%	3.32%	3.38%	2.22%
	RMSE	26.024	23.335	28.166	40.568	39.091	18.873
2017/6/9–2017/7/21	MAPE	2.623%	2.254%	2.108%	3.22%	2.69%	1.18%
	RMSE	27.966	23.180	22.940	30.507	28.785	10.609
2017/9/12–2017/10/23	MAPE	2.66%	2.582%	2.588%	2.89%	2.75%	1.54%
	RMSE	26.9602	22.977	25.710	29.980	26.194	14.390
2017/12/17–2018/1/28	MAPE	2.68%	2.604%	2.152%	3.31%	3.39%	1.49%
	RMSE	28.287	23.138	24.186	40.355	33.092	20.172

where y_i indicates actual value and $\hat{y}_i$ indicates the forecasting value. N indicates the sample size.

Table 2 shows the forecasting errors of the proposed method in this chapter compared with those of other typical machine learning methods (CNN, LSTM, CNN-LSTM, SVM, and random forest). The load forecasting method proposed in this chapter has a small forecasting error for load forecasting in different seasons, especially for summer and winter when the load value is relatively high.

4.4 Discussions

In this section, in order to verify the superiority of the proposed method, the forecasting values of CNN, LSTM, CNN-LSTM and the proposed method (CNN-LSTM with attention mechanism) are compared with the true values, utilizing the electricity load data in summer as an example. The forecasting load values from different forecasting methods and true values are compared in Fig. 4.

(1) The comparison of CNN shows that CNN is stronger than LSTM in terms of fitting, which is because the ability of convolutional kernel of CNN to extract load data features. For example, the convolutional kernel can extract the features of load value changes, but the forecasting is with bad performance at the extreme points of the load curve.
(2) From Fig. 4a, the LSTM forecasts the peak value of the load curve better than that of the CNN. However, the forecasting at the low peak value of the load is not as good as the CNN. LSTM forecasting is slightly smaller than CNN under the criteria of MAPE, which considers more overall features and is also excellent in the overall curve trend.
(3) In Fig. 4c, CNN-LSTM combines the advantages of both CNN and LSTM to a large extent, thus the errors are smaller than a single CNN or LSTM in terms of MAPE and RMSE. However, since the weights of the last implicit layer and the output layer are in the form of random assignment, it fails to highlight the

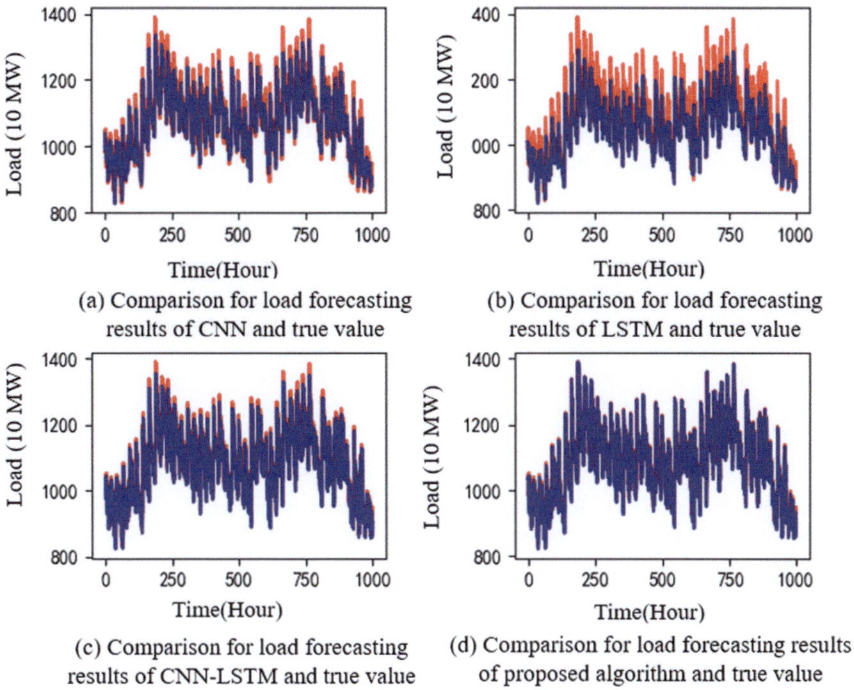

(a) Comparison for load forecasting results of CNN and true value
(b) Comparison for load forecasting results of LSTM and true value
(c) Comparison for load forecasting results of CNN-LSTM and true value
(d) Comparison for load forecasting results of proposed algorithm and true value

Fig. 4 Comparison of forecasting and true load values with different forecasting methods in summer

influence of key information where the extraction for features still needs to be improved.

(4) From Fig. 4d, the proposed load forecasting algorithm is greatly enhanced in fitting and can accurately determine the trend of the true load value, while the forecasting of individual mutation values is also much stronger than the other three methods.

The method proposed in this chapter extracts features from load sequences by CNN, and then uses LSTM with attention mechanism to learn the temporal features in load values to complete the load forecasting. The MAPE of the model is 2.22%, 1.18%, 1.54%, and 1.49% in spring, summer, autumn, and winter, respectively. In the summer with the highest load values, the proposed algorithm reduces MAPE by 1.443%, 1.074%, 0.928%, 2.04%, 1.51%, and RMSE by 17.36, 12.57, 12.33, 19.90, 18.18 compared with CNN, LSTM and CNN-LSTM, SVM, and random forest models, respectively.

5 Conclusion

This chapter considers the load characteristics of DERs in different seasons which are aggregated in VPPs, by dividing the load value data into four categories according to seasonal divisions. Note that it is difficult for CNN to accurately learn data features when the data fluctuates greatly. Moreover, the LSTM may suffer from the phenomenon of lost sequence when the input sequence is too long. The CNN-LSTM cannot decide the weights of the implicit layer of the LSTM. Thus, this chapter proposes an attention mechanism-based CNN-LSTM for short-term load forecasting of DERs aggregated in VPPs with its verification and application prospects.

References

1. X. Meng, F. Gao, T. Xu, K. Zhou, W. Li, Q. Wu, Inverter-data-driven second-level power forecasting for photovoltaic power plant. IEEE Trans. Industr. Electron. **68**(8), 7034–7044 (2021)
2. A. Mnatsakanyan, S.W. Kennedy, A novel demand response model with an application for a virtual power plant. IEEE Trans. Smart Grid **6**(1), 230–237 (2015)
3. A.S. Khwaja, X. Zhang, A. Anpalagan et al., Boosted neural networks for improved short-term electric load forecasting. Electr. Power Syst. Res. **143**, 431–437 (2017)
4. L. Jiang, X. Wang, W. Li, L. Wang, X. Yin, L. Jia, Hybrid multitask multi-information fusion deep learning for household short-term load forecasting. IEEE Trans. Smart Grid **12**(6), 5362–5372 (2021)
5. E. Ceperic, V. Ceperic, A. Baric, A strategy for short-term load forecasting by support vector regression machines. IEEE Trans. Power Syst. **28**(4), 4356–4364 (2013)
6. S.H.R. Nahid-Al-Masood, S.R. Deeba, E. Hossain, A short-term load forecasting method using integrated CNN and LSTM network. IEEE Access **9**, 32436–32448 (2021)
7. Z. Deng, B. Wang, Y. Xu, T. Xu, C. Liu, Z. Zhu, Multi-scale convolutional neural network with time-cognition for multi-step short-term load forecasting. IEEE Access **7**, 88058–88071 (2019)
8. H.H. Goh et al., Multi-convolution feature extraction and recurrent neural network dependent model for short-term load forecasting. IEEE Access **9**, 118528–118540 (2021)
9. J. Li et al., A novel hybrid short-term load forecasting method of smart grid using MLR and LSTM neural network. IEEE Trans. Industr. Inf. **17**(4), 2443–2452 (2021)
10. W. Kong, Z.Y. Dong, Y. Jia, D.J. Hill, Y. Xu, Y. Zhang, Short-term residential load forecasting based on LSTM recurrent neural network. IEEE Trans. Smart Grid **10**(1), 841–851 (2019)
11. M. Alhussein, K. Aurangzeb, S.I. Haider, Hybrid CNN-LSTM model for short-term individual household load forecasting. IEEE Access **8**, 180544–180557 (2020)
12. A. Tascikaraoglu, O. Erdinc, M. Uzunoglu, A. Karakas, An adaptive load dispatching and forecasting strategy for a virtual power plant including renewable energy conversion units. Appl. Energy **119**, 445–453 (2014)
13. W. Kong, Z.Y. Dong, D.J. Hill, F. Luo, Y. Xu, Short-term residential load forecasting based on resident behaviour learning. IEEE Trans. Power Syst. **33**(1), 1087–1088 (2018)
14. Y. Wang, W. Liao, Y. Chang, Gated recurrent unit network-based short-term photovoltaic forecasting. Energies **11**(8), 1–14 (2018)
15. Z. Liao, H. Pan, X. Fan, Y. Zhang, L. Kuang, Multiple wavelet convolutional neural network for short-term load forecasting. IEEE Internet Things J. **8**(12), 9730–9739 (2021)
16. Y. Ju, J. Li, G. Sun, Ultra-short-term photovoltaic power prediction based on self-attention mechanism and multi-task learning. IEEE Access **8**, 44821–44829 (2020)

Chapter 3
Clustering Forecasting of Outputs for VPPs Aggregated with EVs Considering Meteorological Factors

In Chap. 2, the forecasting of VPPs aggregated with various DERs has been introduced. Considering the growth of EVs as a typical kind of DERs, the forecasting of outputs for VPPs aggregated with EVs is presented in this chapter. Accurate forecasting of charging load of EVs has practical significance for the optimal operation of VPPs. Because the characteristics of charging load of EVs are different from traditional electric load, it is necessary to carry out targeted research on the influencing factors and forecasting model of charging load for EVs. Considering the differences of time series characteristics and influencing factors of different types of charging load of EVs, a forecasting model of charging load for EVs considering daily type, maximum and minimum temperatures is established. Fuzzy C-means (FCM) is used to cluster the charging load, where data feature attributes are mined and similar daily load is extracted. The least square support vector machine (LS-SVM) is used to predict the similar daily load after clustering. The forecasting results are compared with the test set, and the results show that the forecasting accuracy based on the proposed model is higher than that of the LS-SVM method, which verifies the effectiveness of the proposed forecasting model.

1 Introduction

As one of the typical resources aggregated in VPPs, EV load is showing a large-scale growth trend, which has an increasingly prominent impact on power systems. Accurate forecasting of EV charging load is conducive to the reliable operation of VPPs, and has become the focus of research and practical application at this stage. For the power grid, accurate load forecasting is the basis for the rational planning and operation of the power grid, which is conducive to the layout of power supply and demand, the realization of effective power supply, and the provision of decision-making basis for the planning management and operation scheduling of transmission and distribution network [1, 2]. EV charging load forecasting can provide support for

H. Jia et al., *Business Models and Reliable Operation of Virtual Power Plants*,
https://doi.org/10.1007/978-981-19-7846-3_3

the aggregation of EVs as VPPs to participate in market transactions, and promote the implementation of demand response of load side resource [3, 4]. Therefore, realizing more accurate EV charging load forecasting has become an important and arduous work to promote the development of VPPs and energy industry.

Generally, there are two ways to improve the accuracy of the charging load forecasting model: increasing the consideration of the influencing factors of the charging load and improving the forecasting algorithm model. Increasing the data input dimension in the charging load forecasting model, fully exploring the external influencing factors affecting the charging load of EVs, such as the service environment and performance of vehicles, and classifying these factors from the aspects of charging capacity, power and time distribution. It can effectively solve the defects of incomplete and inaccurate identification of influencing factors in the traditional model, and ensure accuracy without sacrificing the calculation efficiency of the model [5]. The factors affecting the charging load of EVs mainly include meteorological temperature and traffic conditions [6], as well as date type and driving path, travel time and moving speed affecting user behavior. In the process of forecasting and modeling, the interaction between user behavior and traffic characteristics can be properly considered. User behavior varies greatly under different traffic conditions [7, 8]. At present, the widely used method considering traffic characteristics is to fit the probability distribution of user behavior characteristic parameters of EVs through traditional vehicle traffic statistics [9, 10].

Research has been conducted in load forecasting models, such as artificial intelligence algorithm, and constantly optimized and improved them. Back propagation (BP) neural network is a supervised learning method suitable for multi-layer neural network, which can get better forecasting results, but it has some disadvantages, such as slow convergence speed and prone to local optimization [11, 12]. Support vector machine (SVM) realizes the minimization of structural risk. Its excellent interpretation and generalization ability can effectively solve the problems of small samples and nonlinearity [13]. Least square support vector machine (LS-SVM) is an extension of standard SVM, which can effectively reduce the upper limit of model generalization error and achieve better load forecasting effect in short-term and midterm load forecasting [14, 15]. K-means, hierarchical clustering, spectral clustering and fuzzy C-means (FCM) clustering are commonly used in clustering algorithms. Among them, K-means and FCM are more commonly used in the field of load forecasting. K-means is an exclusive clustering algorithm. FCM extends K-means to fuzzy cases and applies it to fuzzy classification problems [16]. Only the membership degree of data points about each category is calculated and the membership degree weight is provided [17]. Therefore, the application of FCM in the forecasting field is better than the hard-clustering result of K-means.

The charging load of EVs has the same periodicity and fluctuation as the traditional power load, but their load characteristics and influencing factors are not exactly the same. The traditional power load generally presents the state of peak in the daytime and trough at night, while for the charging load of EVs, only the morning has a regular pattern. The charging load of EVs is largely affected by traffic and weather, and due to the diversity of transportation modes, the change elasticity of the use frequency and charging load of EVs is large. At the same time, the EV power battery is more closely related to environmental factors such as temperature. In view of this, it is necessary to carry out targeted research on EV charging load forecasting. In this chapter, the characteristics and influencing factors of EV charging load are briefly analyzed. Considering the influencing factors such as adjacent daily load, date type, weekly attribute and temperature, an EV charging load forecasting model based on cluster analysis and LS-SVM is proposed. The effectiveness of the proposed method is verified by comparing the results of different forecasting methods.

2 Charging Load Characteristics and Influencing Factors of EVs

Before EV charging load forecasting, exploring and analyzing the load characteristics and influencing factors are needed to lay a good foundation for the load forecasting. Figure 1 is the flow chart of analysis for load forecasting.

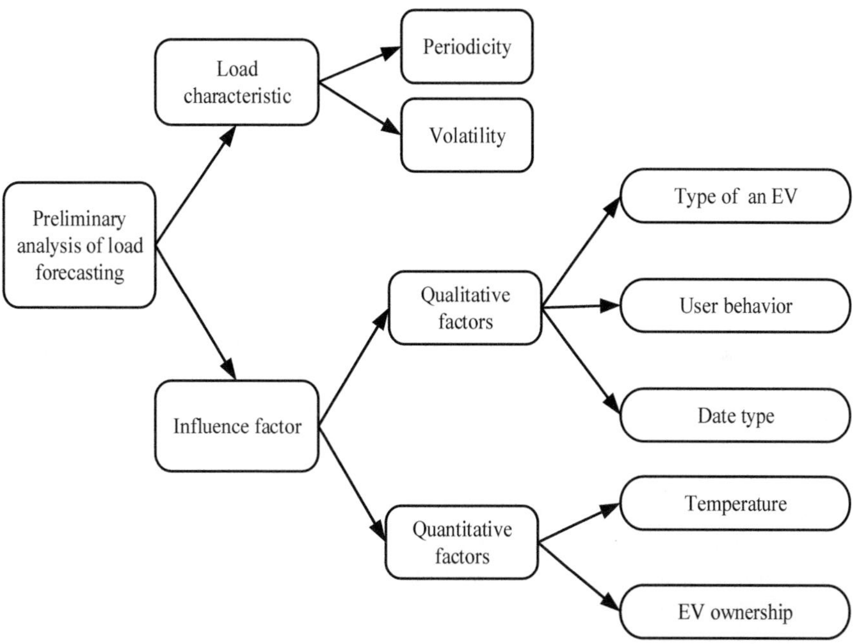

Fig. 1 Preliminary analysis flow of EV charging load

2.1 *Charging Load Characteristics of EV*

(1) Periodicity. The annual periodicity of charging load is reflected in the seasonal periodicity. The charging load in spring and autumn will be significantly lower than that in winter and summer. The periodicity of the week is mainly reflected in the date type. For example, the daily load change of adjacent working days basically follows the same regular pattern. The daily periodicity reflects the load change trend in a day, and the load at the same time of each day shows a roughly similar change trend. The peak load is usually in the noon and afternoon of the day, while the valley load is in the nighttime, and the load change trend in the transition period is between the peak and trough. Daily periodicity is an important basis and analysis basis for short-term and ultra-short-term charging load forecasting.

(2) Volatility. The daily charging behavior will be directly affected by various factors such as the charging time and the type of electric load. In addition, some quantitative indexes can be introduced to quantitatively analyze this characteristic. In this chapter, the index values of daily average load, daily maximum load, daily minimum load and instantaneous average change rate of daily load are obtained through the measured points of charging load, which can reflect the fluctuation characteristics of EV charging load and provide the basis for fuzzy load clustering.

2.1.1 Influencing Factors of EV Charging Load

(1) Type of EVs. EVs can be divided according to different standards, and classification according to vehicle use is a common method. The charging load characteristics of different uses are naturally different [18]. Referring to the classification of traditional fuel vehicles [19], EVs can also be divided into public transport and private transport. Public EVs usually have relatively stable driving distance, charging time and place, and their load characteristics are easy to grasp. Private EVs usually operate flexibly. The driving distance, charging time and location are greatly affected by EV users' behaviors. The randomness of charging power is strong, and the load characteristics are significantly different from public transport.

(2) Date type. The date type mainly affects the charging load of EVs by affecting the travel regular pattern and time of users. Effectively distinguishing the date type is of great significance to improve the forecasting accuracy of the forecasting model [20]. Since the travel regular pattern of users is significantly related to working days and non-working days [21], it is more appropriate to divide the date type into working days and non-working days. Private EVs have the characteristics of personal use. They are usually used as a means of transportation during commuting on weekdays and as a means of transportation for leisure and entertainment activities on non-weekdays. The load characteristics of weekdays and non-weekdays are different. Figure 2 shows the charging load

of working days and non-working days in a typical city from November 2019 to April 2020, with a sampling point for each 15 min. It can be seen from Fig. 2 about the load difference of different date types. During 09:00–22:00, the fluctuation of charging load on working days is significantly greater than that on non-working days.

(3) Meteorological factors. Meteorological factors mainly include environmental temperature, humidity, wind speed, air pressure, etc. [22, 23]. If all meteorological factors are added to the forecasting model, it will increase the redundancy of data information and affect the accuracy and efficiency of calculation. Through the analysis of the charging load data and relevant operation data of the EV charging station, it is concluded that the charging load of the charging station

Fig. 2 Working day and non-working day charging load in a typical city

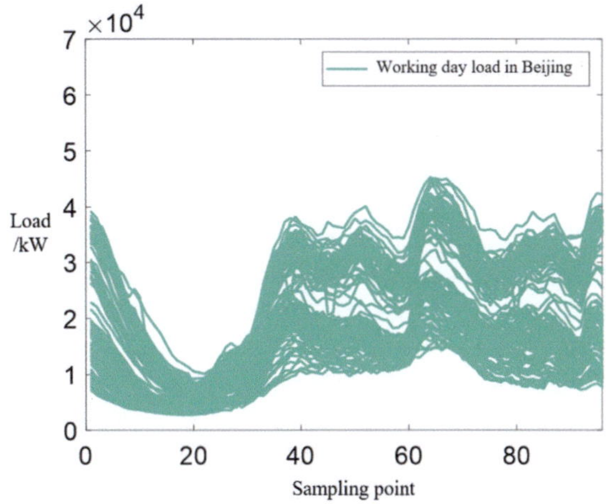

(a) Working day load

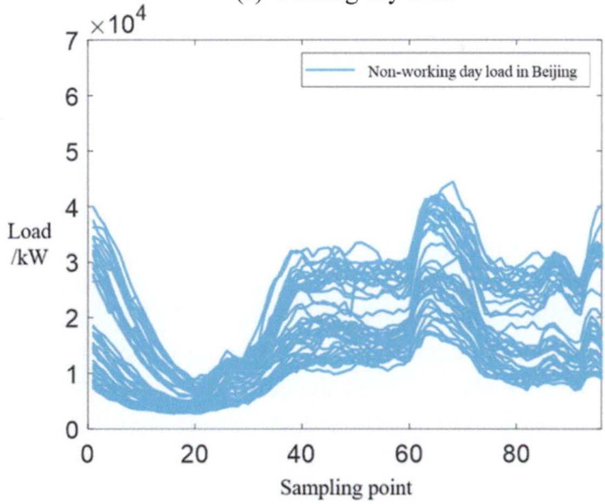

(b) Non-working day load

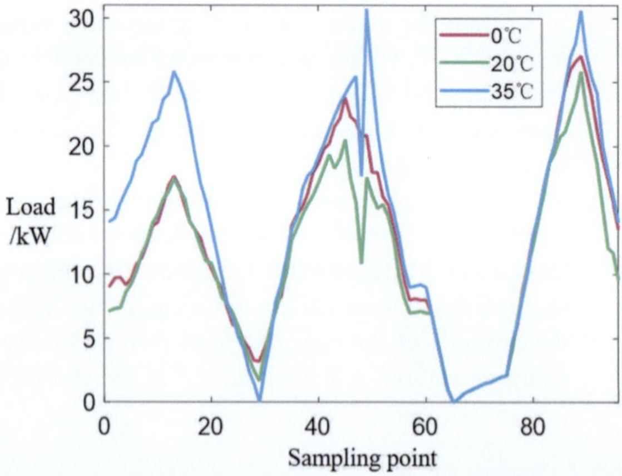

Fig. 3 Typical charging load curve of EVs at different temperatures in a certain area from January to July 2020

has a strong correlation with the temperature and has little relationship with the wind speed and humidity. Therefore, the temperature factors which have a significant effect on the load are considered in this chapter.

EV charging load is more sensitive to meteorological temperature, especially in the peak load period [24]. Figure 3 shows the typical charging load curve of EVs in a region under different temperatures from January to July 2020, with one sampling point every 15 min. It can be seen from Fig. 3 that in the low load period, the influence of temperature on the load is not obvious, and in the peak period, the influence of temperature on the load is significant. Under certain conditions, the peak load and charging capacity increase under the influence of high temperature.

3 EV Charging Load Forecasting Modeling Based on FCM and LS-SVM

Compared with working days, the load data on non-working days are scarce and the regularity of data is difficult to investigate. Therefore, when predicting the charging load of EVs on non-working days, it is necessary to process all historical data and construct a data set with high reference to non-working days. In this chapter, FCM is used to extract the load data with similar load form to the day to be predicted, which can effectively enhance the regularity of the data and make up for the lack of non-working day data. At the same time, considering the external influencing factors and their own change trend, the date type, temperature influencing factor variables and the load at the same time three days before the forecast day are supplemented in the LS-SVM for load forecasting. The process of EV charging load cluster forecasting considering similar daily and meteorological factors is shown in Fig. 4.

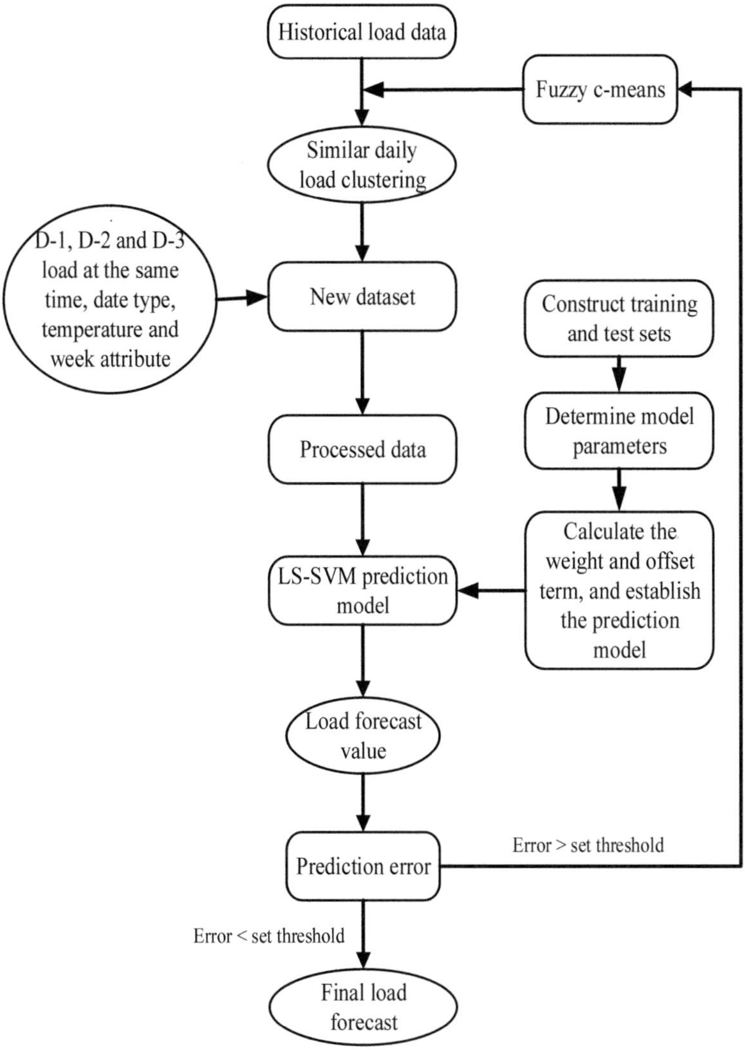

Fig. 4 The flow chart of load forecasting

3.1 FCM Cluster Analysis of Similar Days

Daily load curve usually contains the relevant information of each load characteristic index, and the regularity and continuity of the load curve itself to provide a prerequisite for the effective analysis of load characteristics. Referring to the traditional selection method of power load characteristic quantity, considering that the daily load curve of EVs, this chapter selects the measured data points of daily load curve as the characteristic quantity of fuzzy cluster analysis. The date with similar load curve shape is defined as similar day. By obtaining the minimum value of the Euclidean distance and membership function of similar daily load of the same cluster center, the distance within the similar daily load class is the smallest, and the distance between these classes is the largest. Selecting similar days according to the clustering level can concentrate the effective information of data and speed up the selection speed. The algorithm of FCM is formulated in (1)–(4).

$$X = \{x_1, x_2, ..., x_n\} \in R^s \tag{1}$$

where X is the sample set of a given load; S represents the dimension of samples. N is the number of samples, as well as the number of load curves participating in clustering.

$$\min J(\boldsymbol{U}, \boldsymbol{V}) = \sum_{i=1}^{c}\sum_{j=1}^{n} u_{ij}^{m} d_{ij}^{2} \tag{2}$$

$$\sum_{i=1}^{c} u_{ij} = 1, \quad 1 \le j \le n \tag{3}$$

$$\sum_{j=1}^{n} u_{ij} > 0, \quad 1 \le i \le c \tag{4}$$

where U is the fuzzy partition matrix and V is the cluster center matrix. u_{ij} is the membership degree of sample $x_i$with respect to class j. d_{ij} is the Euclidean distance from the sample x_i to the cluster center of class j. m is the weighted index of fuzziness and c is the number of clusters.

This is an optimization problem with independent variable (U, V) constraints, which is solved by rotation optimization strategy. The general steps of the algorithm are presented as follows.

(1) Initialization: The number of clusters c, the fuzziness weighting index m and the iteration stop limit ε of the data set X are given appropriate value, then, the initial membership matrix $\boldsymbol{U}^{(0)} = \left[u_{ij}^{(0)}\right]_{c \times n}$.
(2) Calculate cluster center v_i:

$$v_i^l = \frac{\sum_{j=1}^{n} (u_{ij}^{(l-1)})^m x_j}{\sum_{j=1}^{n} (u_{ij}^{(l-1)})} \quad (i = 1, 2, ..., c) \tag{5}$$

where l is the current iteration algebra.
(3) Calculate membership u_{ij} and objective function value min $J^{(l)}$:

$$u_{ij}^{l} = \frac{1}{\sum_{k=1}^{c}\left(\frac{d_{ij}^{(l)}}{d_{kj}^{(l)}}\right)^{\frac{2}{m-1}}} (j = 1, 2, \ldots, n) \tag{6}$$

$$\min J^{(l)}(U^{(l)}, V^{(l)}) = \sum_{i=1}^{c}\sum_{j=1}^{n}\left(u_{ij}^{(l)}\right)^{m}\left(d_{ij}^{(l)}\right)^{2} \tag{7}$$

(4) Iterative error judgment: if satisfied

$$\max\left|u_{ij}^{l} - u_{ij}^{l-1}\right| < \varepsilon \tag{8}$$

then the iteration ends, otherwise $l = l + 1$ and returns to step (2) to continue the iteration.

(5) Determine the sample category according to the principle of maximum membership.

3.2 *Load Forecasting Model Based on LS-SVM*

LS-SVM is an improved method of traditional SVM approach, which can be used as a function approximation method for load forecasting. It uses equality constraints to replace the inequality constraints of SVM, transforms the quadratic programming problem into a linear problem, and reduces the computational complexity. LS-SVM is based on statistical theory and uses structural risk minimization to better solve the problems with small samples, nonlinearity and local minima. The algorithm for LS-SVM is demonstrated as follows.

Given the training sample set { (x_i, y_i)} ,i = 1,2,...,n, x_i is the input load value and y_i is the corresponding output value. The regression function is formulated as (9).

$$f(x) = \omega^T \cdot \phi(x) + b \tag{9}$$

where ω is the weight vector and $\varphi(x)$ is the mapping function from low dimensional space to high dimensional space. b is the offset term.

The objective function and constraints are presented in (10)–(11):

$$\min J(\omega, e_i) = \frac{1}{2}\|\omega\|^2 + \frac{1}{2}C\sum_{i=1}^{l} e_i^2 \tag{10}$$

$$y_i = \omega^T \cdot \phi(x_i) + b + e_i, \quad i = 1, 2, ..., l \tag{11}$$

where e_i is the error and C is the penalty parameter. The Lagrange multiplier is introduced as λ, where $\lambda \in R^{|x|}$. The constrained optimization problem can be transformed into an unconstrained optimization problem as (12).

$$L(\omega, b, e_i, \lambda) = \frac{1}{2}\|\omega\|^2 + \frac{1}{2}C\sum_{i=1}^{l} e_i^2 - \sum_{i=1}^{l} \lambda_i[\omega^T \cdot \phi(x_i) + b + e_i] - y_i \quad (12)$$

According to the KKT optimization conditions, the following results are obtained.

$$\begin{cases} \frac{\partial L}{\partial \omega} = 0 \rightarrow \omega = \sum_{i=1}^{l} \lambda_i \phi(x_i) \\ \frac{\partial L}{\partial b} = 0 \rightarrow \sum_{i=1}^{l} \lambda_i = 0, \quad i = 1, 2, ..., l \\ \frac{\partial L}{\partial e_i} = 0 \rightarrow \lambda_i = Ce_i, \quad i = 1, 2, ..., l \\ \frac{\partial L}{\partial \lambda_i} = 0 \rightarrow \omega^{\mathrm{T}} \cdot \phi(x_i) + b + e_i - y_i = 0, \quad i = 1, 2, ..., l \end{cases} \quad (13)$$

Eliminating ω and e_i, the solution of Eq. (13) is presented in (14).

$$\begin{bmatrix} 0 & E^T \\ E & K + C^{-1}I \end{bmatrix} \begin{bmatrix} b \\ \boldsymbol{\lambda} \end{bmatrix} = \begin{bmatrix} 0 \\ Y \end{bmatrix} \quad (14)$$

where $\boldsymbol{E}$ is $[1, 1, \ldots, 1]^{\mathrm{T}}$. $\boldsymbol{I}$ is the identity matrix. $\boldsymbol{\lambda} = [\lambda_1, \lambda_2, \lambda_3, \ldots, \lambda_l]^{\mathrm{T}}$, $\boldsymbol{Y} = [y_1, y_2, y_3, \ldots, y_l]^{\mathrm{T}}$. K is the kernel function mapped to the high-dimensional space through nonlinearity, and the optimal linear regression function of LS-SVM is formulated in (15).

$$f(x) = \sum_{i=1}^{l} \lambda_i K(x, x_i) + b \quad (15)$$

4 Case Study

4.1 Brief Introduction of the Case

In order to verify the effectiveness of load forecasting model embedded with similar day cluster charging proposed in this chapter, the history data of charging load for EVs is adopted. The charging load data of EVs in the area from November 16, 2019 to April 30, 2020 are selected for the simulation. The charging load data at weekends are forecasted with a 15-min interval and a total of 96 time points during a day. Two

Table 1 Comparison of load forecasting accuracy between BP neural network and LS-SVM

Forecast date	λ_{MAPE}/%	
	BP	LS-SVM
2020.4.4	31.61	1.57
2020.4.5	10.10	1.44
2020.4.11	26.49	1.81
2020.4.12	26.22	1.59
2020.4.18	25.91	1.79
2020.4.19	19.91	1.33
2020.4.25	27.34	1.65
2020.4.26	27.81	1.54

commonly used algorithms BP neural network and LS-SVM are used to forecast the load on non-working days in April 2020. The average absolute percentage error (MAPE) is utilized for error analysis. The forecasting errors of the two methods are shown in Table 1 and the load forecasting curve is shown in Fig. 5. The MAPE calculation formula is as formulated in (16), where N is the number of sampling points and i is the sampling point. xi is the actual load value of i sampling point and $x'ixi'$ is the load forecasting value of i sampling point.

$$\lambda_{MAPE} = \frac{1}{N}\sum_{i=1}^{N}\left|\frac{x_i - x_i^{'}}{x_i}\right| \times 100\% \tag{16}$$

x_i x_i'.

Since BP neural network has high sensitivity to the initial value, as well as the randomness between the initial value and the threshold at the beginning of each training, the results and errors of converging to the local optimal solution obtained by selecting different initial values are different. Therefore, this chapter takes the MAPE average value of multiple experiments as the evaluation index of the proposed forecasting method. In the load forecasting of non-working days in April 2020, the MAPE predicted by BP neural network is mainly more than 20%, while the MAPE of LS-SVM is no more than 2%, which fully reflects the advantages of proposed LS-SVM algorithm.

Moreover, by changing the inputs of the load forecasting model, for instance, the utilization of the clustering algorithm as well as the consideration of day types and temperature is regarded as the change conditions. Three scenarios are considered to make load forecasting and comparative analysis.

Scenario 1: a single LS-SVM forecasting model is used, and the input variables are day types, maximum temperature, minimum temperature, and load data of D-1, D-2 and D-3 on non-working days to be predicted at the same time.

Scenario 2: using the forecasting model combined with FCM and LS-SVM, the input variables are the load data of D-1, D-2 and D-3 on non-working days to be predicted at the same time.

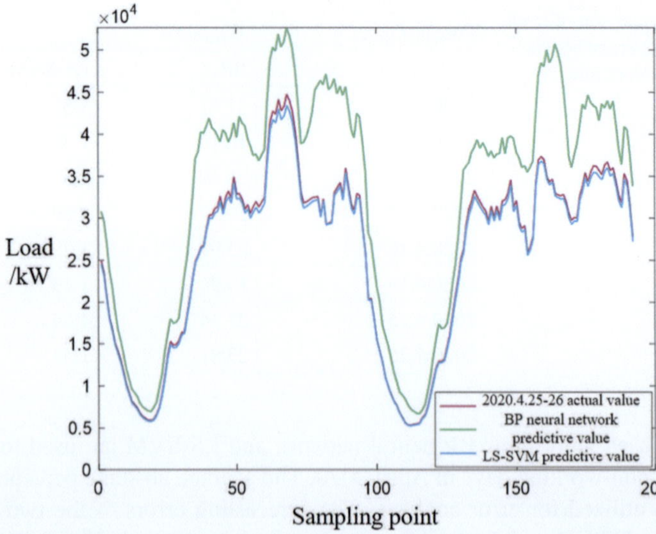

Fig. 5 The comparison of load forecasting results between BP neural network and LS-SVM

Scenario 3: using the forecasting model combined with FCM and LS-SVM, the input variables are date types, maximum temperature, minimum temperature and load data of D-1, D-2 and D-3 on non-working days to be predicted at the same time.

4.2 *Analysis of the Load Forecasting Results*

The load forecasting results and forecasting errors of different scenarios are shown in Fig. 6 and Fig. 7 respectively (April 25–26, 2020 is selected as the forecasting day).

It can be seen from Fig. 6 that there is no significant difference in the forecasting curves of three scenarios. It can be seen from Fig. 7 that the forecasting error of Scenario 3 is significantly lower than that of other methods. The MAPE of the three scenarios are 1.54%, 1.56% and 1.46% respectively. Compared with Scenario 1, Scenario 3 utilizes FCM clustering algorithm to extract similar daily loads on non-working days, and its forecasting error is reduced by 0.08%. Compared with Scenario 2, Scenario 3 considers the influencing factors of load, day types and temperature, thus, its forecasting error is reduced by 0.1%. It can be seen that considering the influence of day types and temperature on charging load and constructing similar data sets on non-working days cannot be ignored to improve the forecasting accuracy.

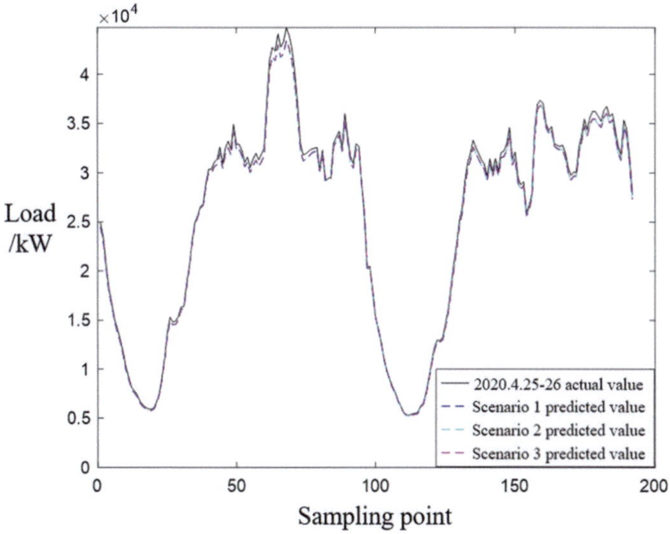

Fig. 6 Forecasting results of different scenarios

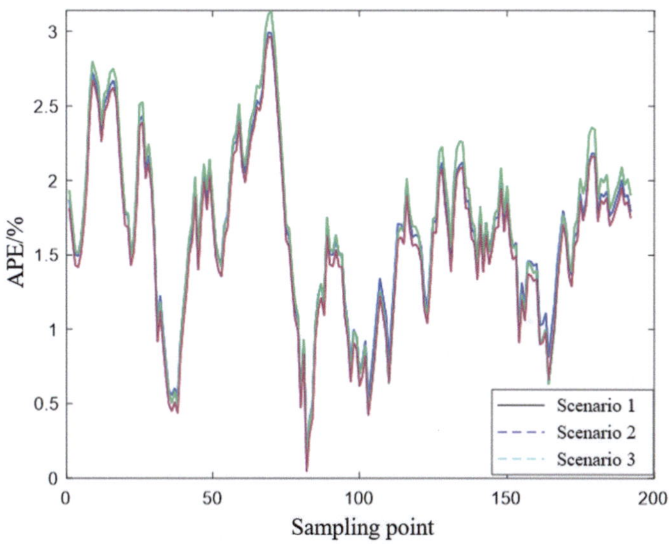

Fig. 7 Comparison of forecasting errors of different scenarios

5 Conclusion

In this chapter, the characteristics and influencing factors of EV charging load are analyzed. According to the characteristics of charging load on non-working day, the characteristic attributes of non-working day load are explored through clustering algorithm. Moreover, the similar day data set of non-working day is extracted. Considering the influence of date type, temperature and historical data, a load forecasting method based on FCM and LS-SVM is proposed. The following conclusions are drawn.

(1) There are lots of complex factors affecting the charging load of EVs. This chapter considers the environmental factors and their own change trend, expands the input variables, and supplements the day types, temperature and the load value at the same time three days before the forecast day. The case results show that the proposed model and method have a good effect on improving the forecasting accuracy.
(2) In view of the lack of data on non-working days, this chapter takes the shape of daily load curve as the fuzzy clustering index, uses FCM for clustering analysis, and makes forecasting on this basis. The case results show that the method has better forecasting accuracy. In other worlds, clustering analysis can effectively improve the forecasting accuracy of charging load on non-working days.

The influencing factors of EV charging load are complex. DERs' behaviors affect the load curve to a great extent, which are diverse and difficult to analyze and forecast. In order to improve the accuracy of load forecasting, further research can be conducted on the influences of DERs' behaviors on load forecasting. Furthermore, based on the accurate load forecasting of DERs, the business models and reliable operation of VPPs can be achieved.

References

1. Z. Liu, D. Wang, H. Jia, N. Djilali, W. Zhang, Aggregation and bidirectional charging power control of plug-in hybrid electric vehicles: generation system adequacy analysis. IEEE Trans. Sustain. Energy **6**(2), 325–335 (2015)
2. M.H.K. Tushar, A.W. Zeineddine, C. Assi, Demand-side management by regulating charging and discharging of the EV, ESS, and utilizing renewable energy. IEEE Trans. Industr. Inf. **14**(1), 117–126 (2018)
3. S. Habib, M.M. Khan, F. Abbas, et al., *Risk Evaluation of Distribution Networks Considering Residential Load Forecasting with Stochastic Modeling of Electric Vehicles, Energy Technology* (2019)
4. G. Fernandez, V. Krishnasamy, J.S.M. Ali et al., Internet of things based real-time electric vehicle load forecasting and charging station recommendation. ISA Trans. **97**, 431–447 (2020)
5. M. Dabbaghjamanesh, A. Moeini, A. Kavousi-Fard, Reinforcement Learning-Based Load Forecasting of Electric Vehicle Charging Station Using Q-Learning Technique. IEEE Trans. Industr. Inf. **17**(6), 4229–4237 (2021)
6. M.B. Arias, S. Bae, Electric vehicle charging demand forecasting model based on big data technologies. Appl. Energy **183**, 327–339 (2016)
7. J. Zhang, J. Yan, Y.Q. Liu et al., Daily EV charging load profiles considering demographics of vehicle users. Appl. Energy **274**, 1–13 (2020)
8. H. Moon, S.Y. Park, C. Jeong et al., Forecasting electricity demand of electric vehicles by analyzing consumers' charging patterns. Transp. Res. Part D: Transp. Environ. **62**, 64–79 (2018)

9. J. Yan, J. Zhang, Y. Liu et al., EV charging load simulation and forecasting considering traffic jam and weather to support integration of renewables and EVs. Renew. Energy **159**, 623–641 (2020)
10. M.S. Islam, N. Mithulananthan, D.Q. Hung, A day-ahead forecasting model for probabilistic EV charging loads at business premises. IEEE Trans. Sustain. Energy **9**(2), 741–753 (2018)
11. S.H. Rafi, Nahid-Al-Masood, S.R. Deeba, E. Hossain, A short-term load forecasting method using integrated CNN and LSTM network. IEEE Access **9**, 32436–32448 (2021)
12. M. Tan, S. Yuan, S. Li, Y. Su, H. Li, F. He, Ultra-short-term industrial power demand forecasting using LSTM based hybrid ensemble learning. IEEE Trans. Power Syst. **35**(4), 2937–2948 (2020)
13. Q. Liu, Y. Shen, L. Wu, J. Li, L. Zhuang, S. Wang, A hybrid FCW-EMD and KF-BA-SVM based model for short-term load forecasting. CSEE J. Power Energy Syst. **4**(2), 226–237 (2018)
14. G. Li, Y. Li, F. Roozitalab, Midterm load forecasting: a multistep approach based on phase space reconstruction and support vector machine. IEEE Syst. J. **14**(4), 4967–4977 (2020)
15. X. Wang, W. Lee, H. Huang, R.L. Szabados, D.Y. Wang, P. Van Olinda, Factors that impact the accuracy of clustering-based load forecasting. IEEE Trans. Industr. Appl. **52**(5), 3625–3630 (2016)
16. J.K. Parker, L.O. Hall, Accelerating fuzzy-c means using an estimated subsample size. IEEE Trans. Fuzzy Syst. **22**(5), 1229–1244 (2014)
17. S.-B. Roh, S.-K. Oh, W. Pedrycz, Z. Wang, Z. Fu, K. Seo, Design of iterative fuzzy radial basis function neural networks based on iterative weighted fuzzy c-means clustering and weighted LSE estimation. IEEE Trans. Fuzzy Syst. **30**(10), 4273–4285 (2022)
18. X. Liu, Bi-level planning method of urban EV charging station considering multiple demand scenarios and multi-type charging piles. J. Energy Storage **48**, 104012 (2022)
19. O. Ghaffarpasand, D.C.S. Beddows, K. Ropkins et al., Real-world assessment of vehicle air pollutant emissions subset by vehicle type, fuel and EURO class: new findings from the recent UK EDAR field campaigns, and implications for emissions restricted zones. Sci. Total Environ. **734**, 139416 (2020)
20. J. Yan, J. Zhang, Y. Liu et al., EV charging load simulation and forecasting considering traffic jam and weather to support the integration of renewables and EVs. Renew. Energy **159**, 623–641 (2020)
21. N. Korolko, Z. Sahinoglu, D. Nikovski, Modeling and forecasting self-similar power load due to EV fast chargers. IEEE Trans. Smart Grid **7**(3), 1620–1629 (2016)
22. R. Jiao, S. Wang, T. Zhang, H. Lu, H. He, B.B. Gupta, Adaptive feature selection and construction for day-ahead load forecasting use deep learning method. IEEE Trans. Netw. Serv. Manage. **18**(4), 4019–4029 (2021)
23. G. Zhang, J. Guo, A novel method for hourly electricity demand forecasting. IEEE Trans. Power Syst. **35**(2), 1351–1363 (2020)
24. C. Wu, S. Jiang, S. Gao, et al., Charging demand forecasting of electric vehicles considering uncertainties in a microgrid. Energy, 123475 (2022)

Chapter 4
Business Models and Practices of Virtual Power Plants

How are virtual power plants making money? How to design business models for VPP operators to achieve sustainable development? As we know, traditional thermal power plants supply capacity when needed and also deliver a range of grid-stabilizing ancillary services, from voltage stabilization to frequency response. VPPs can potentially make money from both types of operations as well.

In this chapter, business models and practices of VPPs are provided. The framework of VPPs participating in electricity market is introduced including VPPs participating in peak shaving energy market, VPPs participating in frequency control market, and VPPs providing comprehensive energy services. Several VPP practices are provided to test and verify the emerging technologies applied in energy industries.

1 Framework of Business Model for VPPs

A business model is simply a plan describing how a business intends to make money. The business model canvas is a strategic management tool that the business idea or concept are visualized and assessed. Both of the traditional business model canvas and the business model for VPPs are provided in this section.

1.1 Traditional Business Model Canvas

The business model canvas was originally developed by Alex Osterwalder and Yves Pigneur as a visual framework for planning, developing and testing the business models of an organization [1]. Business model canvas is a one-page document containing nine boxes that represent different fundamental elements of a business.

H. Jia et al., *Business Models and Reliable Operation of Virtual Power Plants*,
https://doi.org/10.1007/978-981-19-7846-3_4

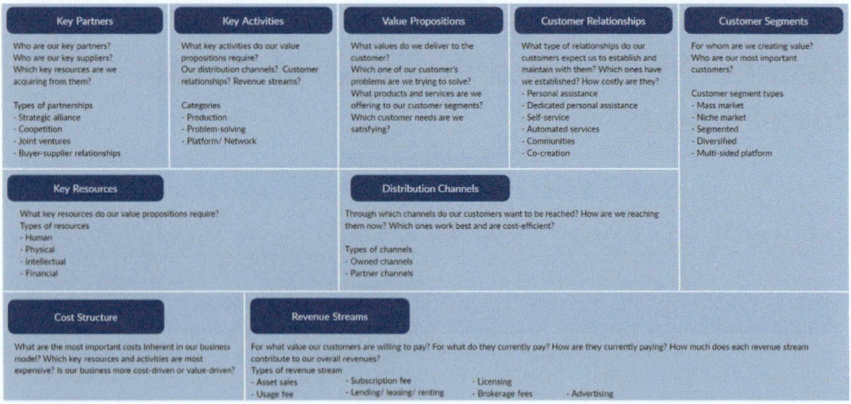

Fig. 1 Business model canvas

There are nine building blocks in the business model canvas: key partners, key activities, value proposition, customer relationships, key resources, distribution channels, customer segments, cost structure, and revenue streams as presented in Fig. 1.

1.2 Business Model for VPPs

With the communication, information, and functional requirement, VPPs can enhance the interaction with DERs and power grids [2]. A large number of market participants with different key resources and profit models is involved in VPPs, including DERs, VPP operators, and market operator. In order to more comprehensively and concisely describe the business model of VPPs, the traditional business model canvas is improved as illustrated in Fig. 2.

In Fig. 2, the key activities including basic activities, extended activities, and value-added activities. For basic activities, VPPs can particate in demand response [3] and electricity market [4] by aggregating DERs. For example, in [5], the VPPs can be aggregated for transient frequency and voltage stability analysis.

For extended activities, VPPs can provide comprehensive energy services for DERs, including energy efficiency management, etc. For value-added activities, VPP can provide energy financial services for DERs, including sharing electricity utilization data of DERs with banks or insurance.

For customer relationships, VPP operators aggregate DERs for electricity trading in electricity market. For value proposition, dispatchable value of VPPs and flexible interactive value among DERs, VPPs and system operator are presented. Key resources are investors, DERs, and ICTs. Cost involves investment cost and operation cost where investment cost includes VPP platform construction cost, terminal equipment construction cost, etc. and operation cost includes operation and maintenance

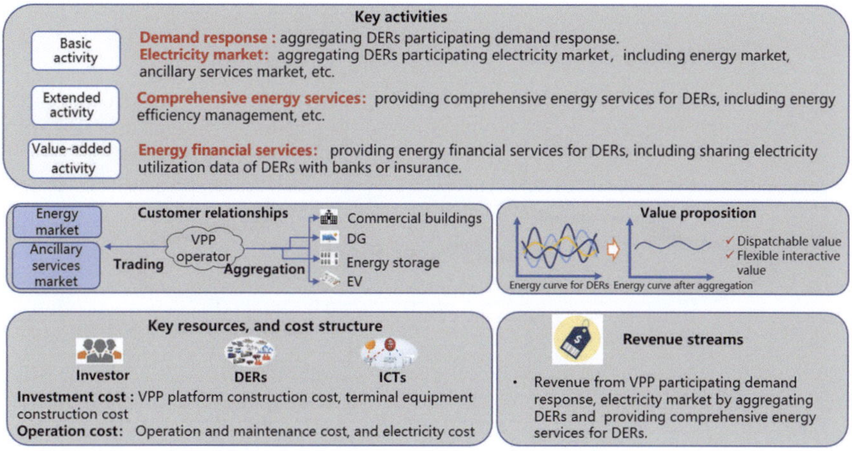

Fig. 2 The business model canvas for VPPs

cost, electricity cost, etc. For revenue streams, revenue is mainly from VPP participating demand response, electricity market by aggregating DERs and providing comprehensive energy services for DERs. Based on the business model canvas for VPPs, typical business model for VPPs participating in electricity market can be demonstrated in the following section.

2 VPPs Participating in Electricity Market

The framework for VPPs participating in electricity market is illustrated in Fig. 3. VPP operators can aggregate various DERs to participate in electricity market with certain market mechanism and operation model, for example, energy market, ancillary service market. VPP operators distribute benefits from electricity market to DERs. Moreover, VPP operators can provide comprehensive energy services for DERs to get revenues. In the following sections, the detailed explanations of VPPs participating in peak shaving energy market, participating in frequency control market, providing comprehensive energy services are provided.

2.1 Normal Mode and Emergency Mode of VPPs in Power System Operation

When VPP operators participate in electricity market, it is essential to know how VPPs participate power system operation. The normal mode and emergency mode of VPPs in power system operation are illustrated in Figs. 4 and 5, respectively.

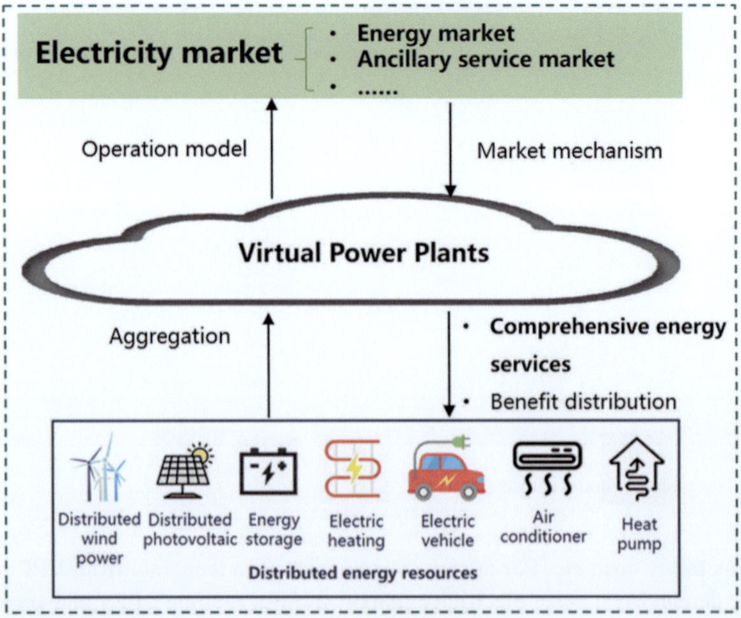

Fig. 3 The framework for VPPs participating in electricity market

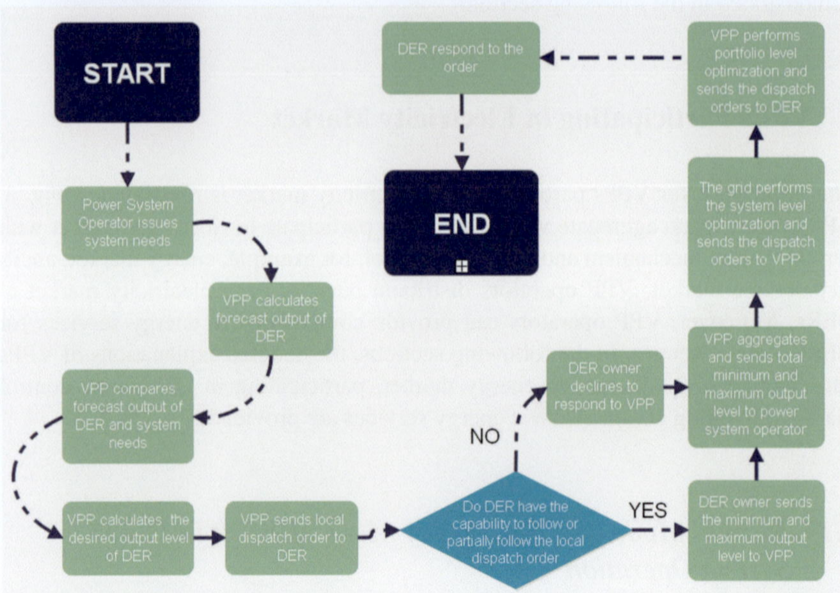

Fig. 4 Normal mode of VPPs in power system operation

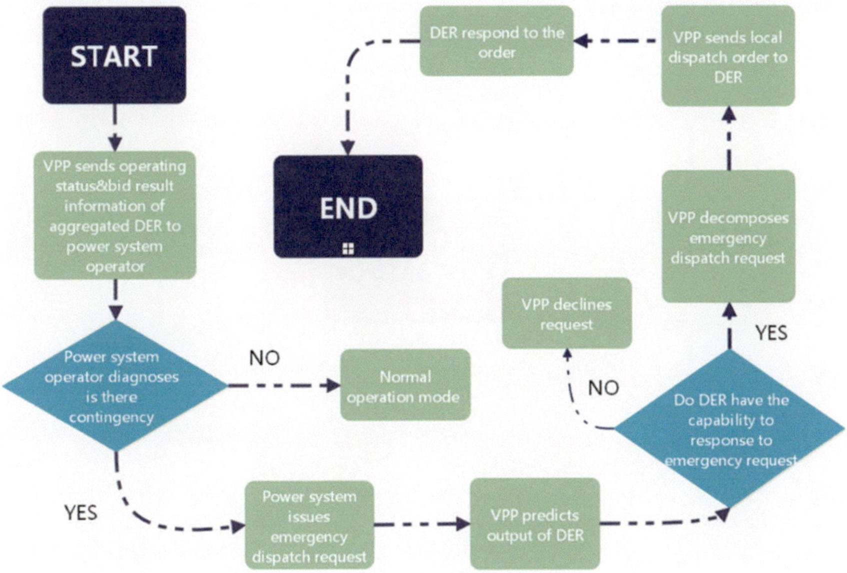

Fig. 5 Emergency mode of VPPs in power system operation

It can be seen from Fig. 4 that for the normal mode of VPPs in power system operation, VPP operators compare forecast output of DERs and system need and send local dispatch order to DERs. If DERs have the capacity to follow or partically follow the local dispatch order, DER owners send the minimum and maximum output levels to VPP operator. If not, DER owners decline to respond to VPP operator. Then, VPP operators aggregate and send total minimum and maximum output levels to power system operator. The power system operator performs the system level optimization and sends dispatch order to VPP operators. Then, VPP operators perform portfolio level optimization and send control signals to DERs with the response of DERs to control signals.

It can be seen from Fig. 5 that for the emergency mode of VPPs in power system operation, VPP operators send operating status and bid result information of aggregated DERs to power system operators. If a power system is in a contingency state, the power system operator will issue emergency dispatch request. Then, VPP operators predict their power outs of DERs. If DERs have the capability to response to emergency request, VPP operators decompose emergency dispatch request and DERs respond to the order sent by VPP operators. If DERs have not the capability to response to emergency request, VPP operators will decline the request from the system operator.

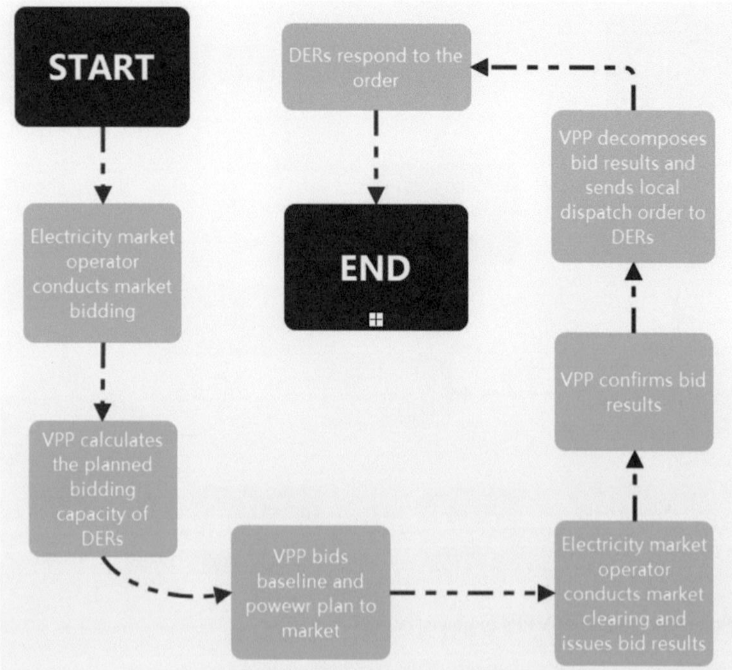

Fig. 6 The procedures of VPPs participating peak shaving energy market

2.2 *VPPs Participating in Peak Shaving Energy Market*

In this section, the procedures of VPPs participating in peak shaving energy market is demonstrated in Fig. 6. With the marking bidding conducted by electricity market operator, VPP operators bid baseline and power plan to electricity market. Then, electricity market operator conducts market clearing and issues bid results. With the conformation of bid results from VPP operators, VPP operators decompose bid results and send local dispatch order to DERs. Finally, DERs make response to local dispatch order ending the procedures of VPPs participating in peak shaving energy market.

2.3 *VPPs Participating in Frequency Control Market*

For the operation of power systems, it is essential to be close to the nominal frequency, where fast controlling reserves is able to decelerate frequency deviations due to disturbances. The stochastic characteristics of both power consumption and power integration of renewable energies bring variations into the system frequency [6]. To maintain the system frequency, both of the primary frequency control service and secondary frequency control service are provided in electricity market scheduling [7]. A large frequency deviation caused by a sudden wind or solar drop-down can be arrested and then stabilized to a new steady-state level by the primary frequency control [8]. The error between the resulting steady-state and the nominal frequency is corrected by the deployment of the secondary frequency control [9] typically deployed by automatic generation control (AGC). Since the characteristics of VPPs can represent as traditional power plants, VPPs have the ability to provide frequency control services including primary frequency control service and secondary frequency control service in frequency control market. In this section, the procedures for VPPs participating in primary frequency control ancillary service market and secondary frequency control ancillary service market are illustrated in Figs. 7 and 8, respectively.

With the submission of parameters for VPPs participating in primary frequency control ancillary service market to the system operator, if VPPs have the capability to provide primary frequency control service, corresponding regulatory authority checks the actual performance provided by VPPs, and end the frequency control service.

For the VPPs participating in secondary frequency control ancillary service market, after the power system operator publish system secondary frequency control needs, VPP operators forecast outputs of aggregated DERs and then bid in the secondary frequency control service market. After the clearing of day-ahead market, the power system operator sends frequency control capacity to VPP operators. Then,

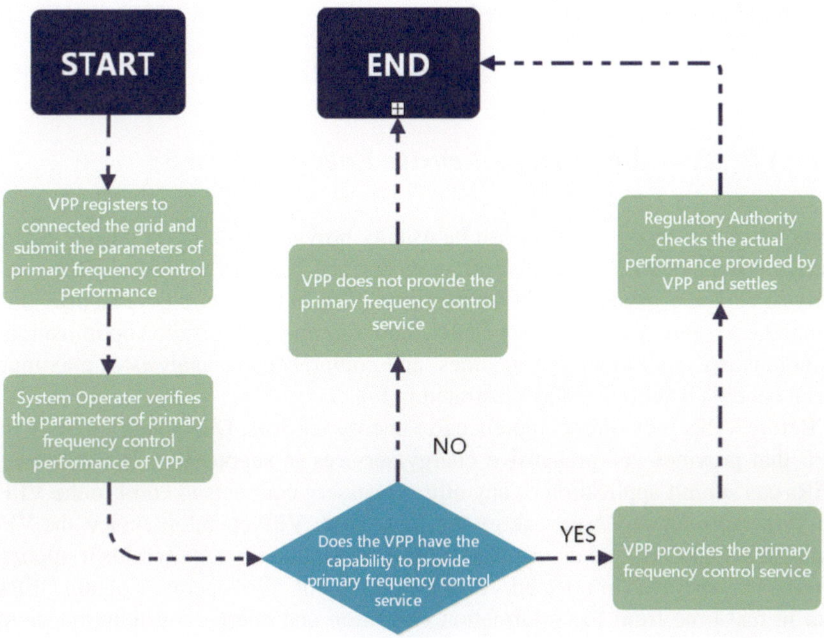

Fig. 7 The procedures for VPPs participating in primary frequency control ancillary service market

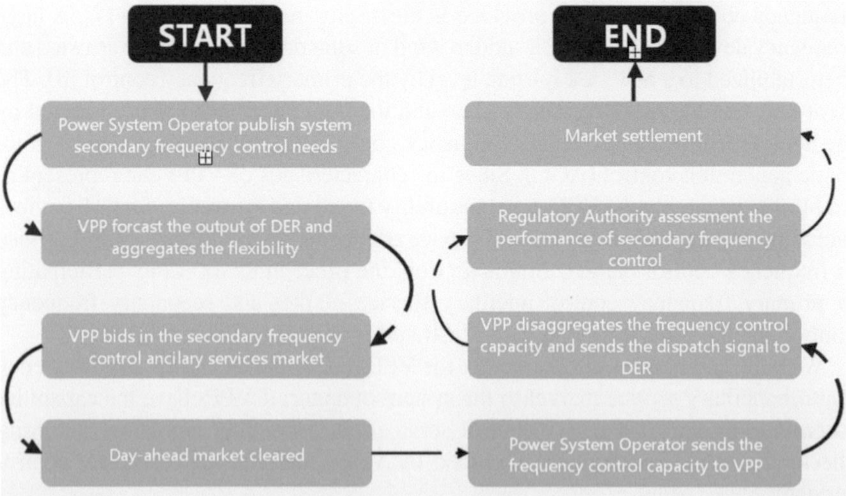

Fig. 8 The procedures for VPPs participating in secondary frequency control ancillary service market

VPP operators disaggregate the frequency control capacity and send dispatch signal to DERs. After the regulatory authority checks the actual performance provided by VPPs for secondary frequency control, the settlement between the electricity market operator and VPP operators, as well as the settlement between VPP operators and DER owners are completed.

2.4 VPP Providing Comprehensive Energy Services

Comprehensive energy services can be usually provided by comprehensive energy service providers with innovative services for end-users as to improve the flexibility of both energy supply and demand side [10]. In this section, VPPs providing comprehensive energy services for DERs, including energy utilization optimizations, economic analysis to increase incomes, and comprehensive analysis to maximum social benefits is introduced as illustrated in Fig. 9.

Before VPPs providing comprehensive energy services, DERs' enrolment in the VPP that provides comprehensive energy services is supposed to be completed. DERs can submit application or any other necessary contracts to enrol in the VPPs that provides comprehensive energy services. Then, VPP operators review the VPP applications and respond as whether further information is required, or if applications can be approved or rejected. After the enrolment, VPP operators obtain DERs' data in real-time from the information collection and energy consumption monitoring system. Through the analysis of energy utilization characteristics and individual energy demand for DERs, VPP operators formulate DERs' models. Then,

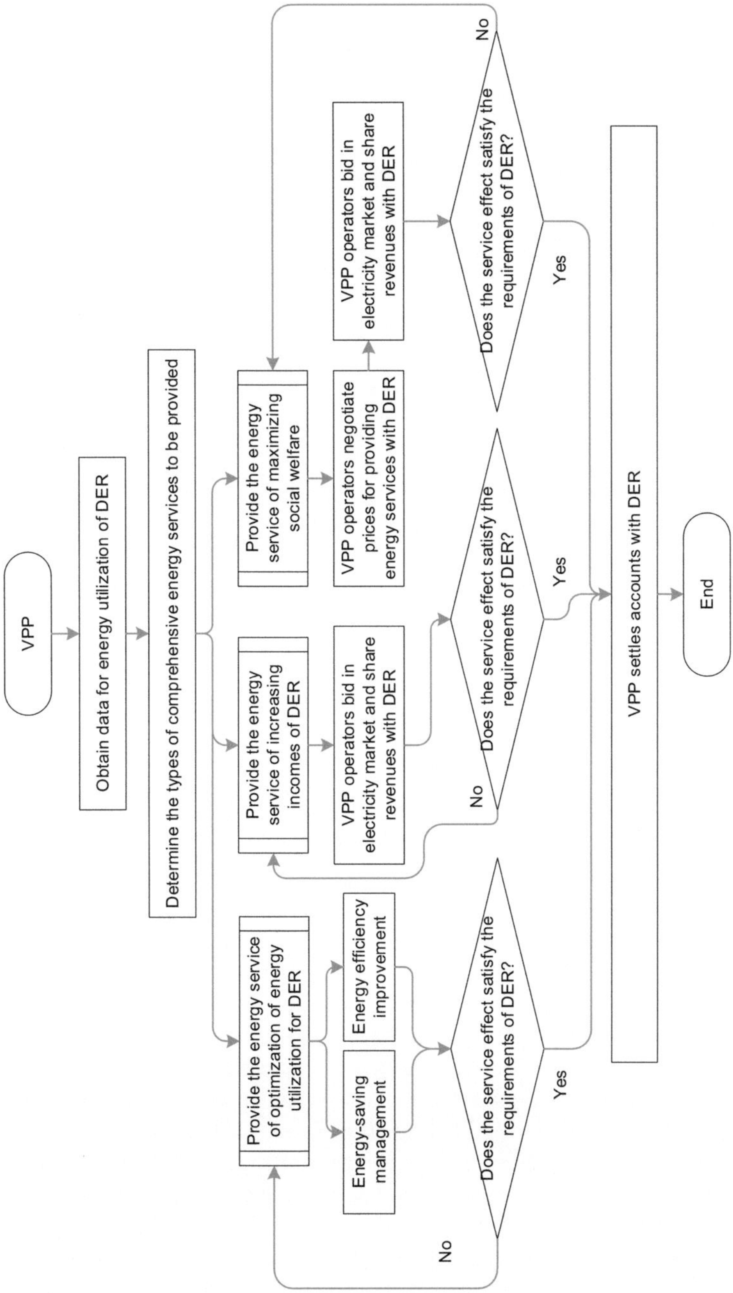

Fig. 9 The procedures for VPPs providing comprehensive energy services

VPP operators determine the types of comprehensive energy services for different DERs, including energy utilization optimization, increasing revenue of DER, and maximizing social benefits. For energy utilization optimization, VPP operators can provide the comprehensive energy services including energy-saving management and energy efficiency improvement to DERs. For increasing revenue of DERs, VPP operators aggregate DERs participating in the electricity market including ancillary service market, etc. Then, VPP operators share the revenues with DERs according to their contributions in the electricity transaction. For maximizing social benefits, VPP operators provide the comprehensive energy services of maximizing social benefits. If the service effects satisfy the requirements of DERs, VPP operators setell account with DERs and end the procedures of VPPs providing comprehensive energy services.

3 VPP Applications in Practice

With the business models for VPPs, VPPs are still being explored and developed all over the world [11]. For example, Bosch, and other well-known companies have developed their respective platforms to support the operation of VPPs [12, 13]. In China, VPP applications in Shanghai and North Hebei have also launched the system construction for VPP operation [14, 15].

3.1 The VPP Practice of Bosch

The VPP practice of Bosch was conducted the Rheintal region of Vorarlberg, Austria [12]. The aggregated DERs in the VPP have been photovoltaic generations, power loads and electric vehicles. In the future, energy storage will be integrated to improve the overall performance. The more facilities are connected to the VPP, the more flexibly and efficiently it can operate. At the same time, however, this networking makes managing the system more complex. For the desired power to be supplied as needed, the system must ensure the seamless interaction of the decentralized sources of power. It must also be able to make reliable forecasts on the capacity of individual power plants. Against this backdrop, Bosch has developed software that continuously determines ideal production volumes and makes recommendations with regard to the management of power plants. In addition to drawing on weather forecasts, the "VPP Manager" also takes price developments on energy markets into account as presented in Fig. 10 [16]. In so doing, the system can determine the best point in time to sell excess power. The primary objective of the VPP is to optimize the aggregated DERs to balance local loads and shave peak loads to delay power grid construction. DERs' electricity consumption data is collected every 15 min by the VPP operator to forecast the further load curve.

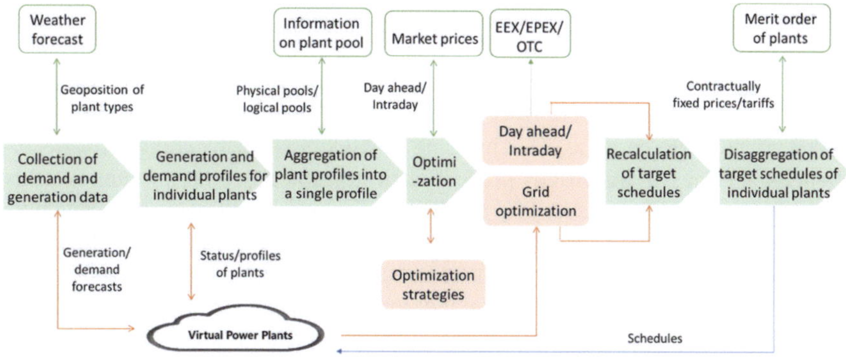

Fig. 10 Optimization of the VPP manager

3.2 *VPP Practice of Next Kraftwerke*

Next Kraftwerke is one of the largest renewable energy power traders in Europe, and operates in Germany to provide various VPP services. The platform for the VPP is called NEMOCS as presented in Fig. 11 [13]. The VPP not only allows to aggregate thousands of electricity producers, consumers, and storage units, but also enables offering their power and flexibility into different markets by intelligently controlling their feed-in and consumption. As the operator of one of the largest VPPs in Europe, the VPP of Next Kraftwerke network thousands of DER assets. Until Q4/2021, the networked capacity is 9.834 MW and the volume of traded energy is 15.1 TWh.

A wide range of business areas for distributed generation operators, load aggregators, utility companies, grid operators, and energy traders is provided by NEMCOS

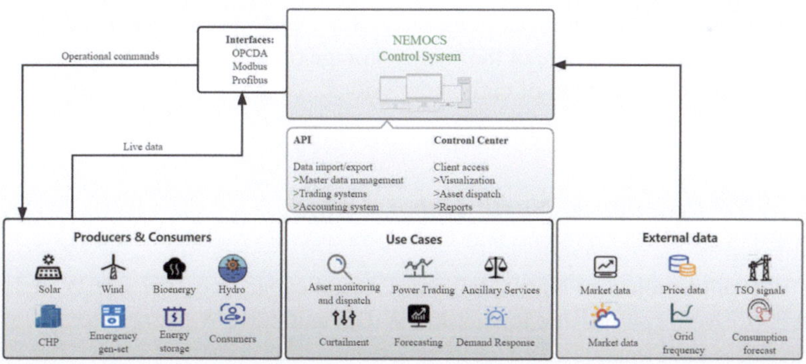

Fig. 11 Communication and data flow of the VPP of next Kraftwerke (https://www.next-kraftwerke.com/)

including energy trading, dispatch, balancing services, curtailment, load management & demand response. For the energy trading, a great amount of different energy sources can be bundled to achieve the necessary size to participate in power and balancing energy markets. For dispatch, scheduling flexible assets based on current market prices can be easily realized by using NEMOCS as a dispatching software. For balancing services, the electricity production of steerable assets at short notice can be increased or reduced to ensure grid stability and achieve additional revenues on the balancing energy markets. For curtailment, the aggregated DERs of the VPP can be shut down within seconds via the control system of the VPP in case of negative electricity prices or imbalances in portfolio. For load management and demand response, the VPP can aggregates commercial and industrial consumers to control electricity consumption based on price signals from the power markets, offer balancing services or control flexible loads.

3.3 VPP Practice in Shanghai, China

By aggregating buildings, DGs, EVs, CCHPs, energy storage facilities, charging piles, and other demand-side resources, the VPP operated in Shanghai, China, mainly participates in the demand response market [15].

On May, 2021, a large-scale use of the VPP in Shanghai for carbon reduction was conducted by aggregating tens of thousands of DERs including lighting, refrigeration, etc. Through remote management and accurate adjustment of the power consumption of these DERs, the load of the whole power grid increased by 412 MW. During two days, the load of the power grid was adjusted by 562 MW and accommodated 1236 MWh of clean energy. This demand response action accurately aggregates 11 kinds of DERs, such as industrial load, commercial buildings, micro grid, distributed energy, CHP, energy storage facilities, ice storage, public charging stations, residential charging piles, local base stations, etc., with a wide range of DER categories. It has realized the full coverage of the energy storage of the local base stations and also contributed to the reduction of carbon emissions.

3.4 VPP Practice in North Hebei, China

The VPP pilot project in North Hebei has a total capacity of 357.6 MW including both 356.7 MW controllable load, 106 kW DG and 800 kW energy storage, where controllable load includes 73.2 MW industrial and commercial loads, 259.7 MW electric heating devices, and 23.8 MW other types of controllable load. The project covers multiple regions in the power system of North Hebei including Zhangjiakou, Langfang and Qinhuangdao cities [15].

The goal of the VPP project is to provide peak shaving services for power system during peak/valley periods of load profiles. With the dispatchable DERs aggregated

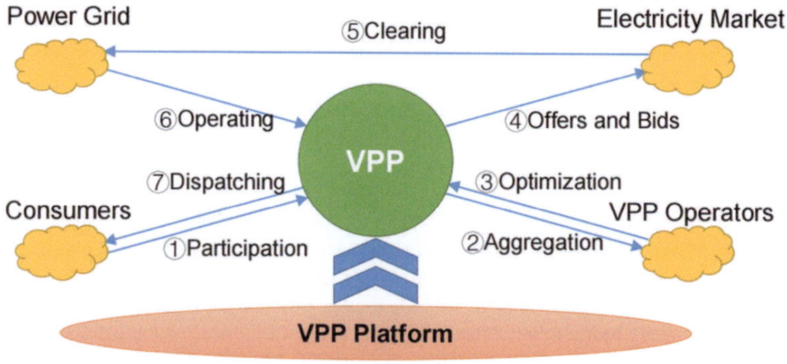

Fig. 12 The operation flow of VPP in North Hebei, China

by the VPP, traditional thermal power plants can reduce the needs for peak shaving and thus save operational costs. Local DERs can be aggregated to balance renewable energy generation and random load. The operation flow of the VPP in North Hebei, China is illustrated in Fig. 12, including participating, aggregation, optimization, offers and bids, clearing operating, and dispatch. From Dec. 2019 to Apr. 2020, the VPP pilot project in North Hebei participated in peak shaving auxiliary service market with the amounts of total regulation 7853.3 MWh and total revenue ¥1.60 million.

References

1. A. Osterwalder, Y. Pigneur, *Business Model Generation* (John Wiley and Sons, 2010)
2. N. Etherden, V. Vyatkin, M.H.J. Bollen, Virtual power plant for grid services using IEC 61850. IEEE Trans. Industr. Inf. **12**(1), 437–447 (2016). https://doi.org/10.1109/TII.2015.2414354
3. Y. Wang, X. Ai, Z. Tan, L. Yan, S. Liu, Interactive dispatch modes and bidding strategy of multiple virtual power plants based on demand response and game theory. IEEE Trans. Smart Grid **7**(1), 510–519 (2016)
4. L. Baringo, M. Rahimiyan, *Virtual Power Plants and Electricity Markets: Decision Making Under Uncertainty* (Springer, 2020)
5. J. Chen, M. Liu, F. Milano, Aggregated model of virtual power plants for transient frequency and voltage stability analysis. IEEE Trans. Power Syst. **36**(5), 4366–4375 (2021). https://doi.org/10.1109/TPWRS.2021.3063280
6. M. Perninge, R. Eriksson, Frequency control in power systems based on a regulating market. IEEE Trans. Control Syst. Technol. **26**(1), 27–37 (2018)
7. G. Zhang, E. Ela, Q. Wang, Market scheduling and pricing for primary and secondary frequency reserve. IEEE Trans. Power Syst. **34**(4), 2914–2924 (2019)
8. Y. Yang, J.C.-H. Peng, Z.-S. Ye, A market clearing mechanism considering primary frequency response rate. IEEE Trans. Power Syst. **36**(6), 5952–5955 (2021)
9. K. Jiang, H. Su, H. Lin, K. He, H. Zeng, Y. Che, A practical secondary frequency control strategy for virtual synchronous generator. IEEE Trans. Smart Grid **11**(3), 2734–2736 (2020)
10. Y. Wu, Y. Wu, J.M. Guerrero, J.C.Vasquez, A comprehensive overview of framework for developing sustainable energy internet: from things-based energy network to services-based management system. Renew. Sustain. Energy Rev. **150**, 111409 (2021)

11. T. Mao et al., Virtual power plant platforms and their applications in practice: a brief review. IEEE Sustain. Power Energy Conf. (iSPEC) **2020**, 2071–2076 (2020). https://doi.org/10.1109/iSPEC50848.2020.9351147
12. Field report: A virtual power plant for the Smart City Rheintal [EB/OL] (2019). https://blog.bosch-si.com/smart-grid/field-report-a-virtualpower-plant-for-the-smart-city-rheintal/
13. VPP Software as a Service NEMOCS—Our Platform for your Virtual Power Plant [EB/OL]. https://www.next-kraftwerke.com/products/vppsolution
14. L. Ya, Z. Deliang, W. Xuanyuan, A peak regulation ancillary service optimal dispatch method of virtual power plant based on reinforcement learning, in *IEEE Innovative Smart Grid Technologies—Asia (ISGT Asia)* (Chengdu, China, 2019), pp. 4356–4361
15. Y. Han, H. Jia, D. Liu, X. Wang, Z. Liu, H. Zhang, Practice of virtual power plants participating in electricity markets in China, in *2021 IEEE 4th International Electrical and Energy Conference (CIEEC)* (2021), pp. 1–5. https://doi.org/10.1109/CIEEC50170.2021.9510325
16. https://www.bosch.com/stories/technology-for-the-energy-turnaround/

Chapter 5
Integrated Electricity/Heat Demand Response for Virtual Power Plants

With the rapid development of DERs and energy storage, the demand for multiple types of loads such as electricity and heat in VPPs continues to grow, which poses an even greater challenge for optimizing VPP operations considering comfort of consumers. In this regard, an integrated electricity/heat demand response (DR) modeling approach for VPPs with comfort of consumers is proposed. In this chapter, by describing the components in an electric/thermal VPP, the power supply output model and the load DR model based on the peak and valley time tariffs are established. The operating cost optimization model of VPPs with the consideration of comfort constraints is proposed based on average thermal sensitivities predictors. The operating costs of VPPs under different DR implementation scenarios through practical examples are compared and analyzed the to guide the selection of the optimal integrated electricity/heat DR solutions for VPPs. It is verified that the integrated DR for electric and thermal loads helps to mitigate the impact of high load demand on the operational stability of VPPs and effectively optimizes the operating costs of VPPs.

1 Introduction

With the development of DERs and energy storage, the electricity consumption of consumers is growing rapidly, where the demand for electricity, heat and other loads in the distribution network continues to increase. If the load exceeds the allowable limit during peak hours, it will definitely threaten the safe and stable operation of the distribution network [1]. VPPs can aggregate multiple DERs and participate in the electricity market as a whole with the continuous growth in demand for multiple types of loads. Achieving optimal scheduling of multiple loads and meeting customer comfort are major challenges for the operation optimization of VPPs. DR refers to the behavior of electricity consumers who actively change their original electricity consumption patterns in response to market tariff signals or incentives. Reference [2] proposed the concept of DR in a VPP considering the uncertainty of response and

H. Jia et al., *Business Models and Reliable Operation of Virtual Power Plants*,
https://doi.org/10.1007/978-981-19-7846-3_5

established the DR model with uncertainties of VPPs and traditional generating units to participate in the optimization model of system scheduling together. A stochastic scheduling model for VPPs considering uncertainties, price-based DR and incentive-based DR was established in [3] to reduce the impact of wind and solar PV on the uncertainty of VPP operation based on robust optimization theory. In [4], the optimal bidding strategy problem of a VPP participating in the day-ahead, real-time and spinning reserve markets was addressed maximizing the VPP profit considering capability from both supply and demandsides for providing reserve services. In [5], a new perspective in the simultaneous use of DR loads and available electric vehicles was introduced in parking lots toward energy management of industrial VPPs aiming at augmenting the system's profits as well as increasing the grid reliability under peak load conditions and cutting down on industrial centers load shedding. A DR energy management model was established in [6] for VPPs considering risk constraints to reduce the operating costs of VPPs while ensuring system power quality. In [7], a data mining-driven incentive-based DR scheme to simulate electricity transactions between VPPs and their participants was proposed which effectively reduced the customer's electricity load while maximizing the use of distributed energy.

However, the DR of a single kind of load can no longer meet the continuous growth of demand for multiple types of loads [8]. For the joint scheduling of multiple types of loads and integrated DR, in [9, 10], the coordinated operation mechanism of integrated DR energy under multiple energy markets. A day-ahead integrated energy system optimal dispatching method considering power to gas units and dynamic pipeline networks was proposed in [11] where a cooperative dispatching strategy for P2G and pipeline storage capability was presented to catch the flexibility. In [12], an online algorithm for energy storage management in networked microgrids with co-generation based on the concept of quality-of-service in electricity was proposed. The concept of networked microgrids with distributed renewable energy supply and co-generation made power supply smarter for electricity/heat using, which had advantages of increasing power supply efficiency and reliability by coordinately scheduling the power supply in a networked way. In [13] a rural multi-energy complementary system structure was presented, which established the output model of wind power, biogas cogeneration, firewood-saving stoves, photovoltaic heat collectors, and air source heat pumps. The combined heat and power economic dispatch was an attractive and momentous optimization problem where the major objective was to find an optimal generation schedule of heat and power to meet the heat and power demands with the minimum cost where an adaptive cuckoo search with differential evolution mutation was put forward in [14].

Further application of integrated DR technology for multiple types of loads to VPPs can be found. In [15], the addition of heat pumps and thermal storage DR was analyzed in different scenarios to minimize the annual costs, fuel consumption, and carbon emissions. An optimal scheduling model for VPPs considering electric and thermal DR for the coordinated scheduling of distributed energy sources such as controllable units, energy storage devices, and wind turbines was proposed [16–18].

However, customer comfort of the integrated DR of multiple loads is not considered in the above literatures, which can directly affect the motivation of customers' response to participate in VPPs and thus affect the operation of VPPs. Reference [19] investigated the joint optimization problem of EVs and household energy dispatch to reduce household electricity costs while considering user comfort preferences. A multi-objective mixed integer nonlinear programming model was proposed in [20] to optimize different and conflicting objectives including energy cost, and user comfort where the effect of different selling prices on the objectives was considered. In [21], an optimal dispatching model of the distribution network was proposed to improve the system flexibility while ensuring user comfort and operation economy. Reference [22] fully considered the comfort of customers in VPPs where the heat and electricity loads can be flexibly adjusted for effectively alleviating the phenomenon of wind abandonment.

In summary, this chapter focuses on load shifting and reduction by implementing price-based DR according to time-of-peak and valley tariffs and incentive-based DR according to the predicted mean vote (PMV) for electric and thermal loads, respectively, taking into account customer comfort. Then, with the objective of minimizing the operating cost, a VPP cost optimization model is established. The operating costs of VPPs in different DR scenarios and impacts of user comfort on the operating costs of VPPs are compared to verify the validity of the proposed model.

2 Operation Principles of a VPP with DR

As illustrated in Fig. 1, a VPP in this chapter includes wind power plant (WPP), photovoltaic (PV), storage cell (SC), heat storage tank (HS), electric boiler (EB) with electric load (EL) and heat load (HL).

The operation framework of a VPP is shown in Fig. 1, which mainly includes three parts: the energy supply side, the customer side and the grid side. Through the energy supply side of the wind turbines and photovoltaic equipment for power

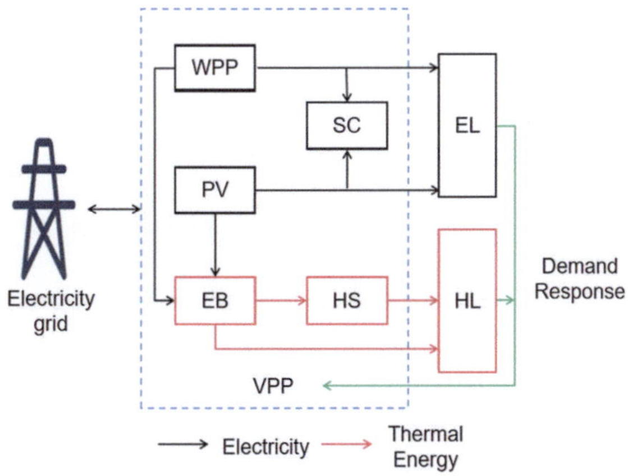

Fig. 1 The operation framework of a VPP

supply, the VPP control center can implement price based DR to electrical load (EL), in order to achieve the peak and valley load shaving. At the same time through the electric boiler, electricity can be transferred into heat. When the power supply is insufficient, the battery will first assist in supplying electricity. If the heating is insufficient, then the appropriate incentive-based DR will be implemented for the thermal load to motivate customers to actively participate in the response, while the heat storage tank will assist in heating. If the power demand still exceeds the supply, electricity will be bought by a VPP from the power grid. When the power supply is sufficient, the surplus power will be preferentially stored and then sold to the power grid to reduce the risk of higher demand for electricity and heat to the operation of a VPP.

3 Output Model and Integrated DR Model

3.1 *Power Supply Equipment Output and Price-Based Electric Load DR Model*

The power supply equipment in a VPP in this chapter mainly includes wind power, photovoltaic power and battery.

(1) WPP output model:

$$g_{\mathrm{WPP}}(t) = \begin{cases} 0\,, v_t \leq v_{\mathrm{in}} \\ \dfrac{v_t - v_{\mathrm{in}}}{v_{\mathrm{R}} - v_{\mathrm{in}}} g\,, v_{\mathrm{in}} \leq v_t \leq v_{\mathrm{R}} \\ g\,, v_{\mathrm{R}} \leq v_t \leq v_{\mathrm{out}} \\ 0\,, v_t \geq v_{\mathrm{out}} \end{cases} \tag{1}$$

In the formula, $g_{\mathrm{wpp}}(t)$ is the actual output of a WPP. g is rated power of a WPP. v_{in} and v_{R} are the cut-in wind speed and rated wind speed. v_{out} and v_t are the cut-out wind speed and the real-time wind speed.

(2) PV output model:

$$g_{\mathrm{PV}}(t) = \eta_{\mathrm{PV}} S_{\mathrm{PV}} \theta(t) \tag{2}$$

In the formula, η_{PV} is solar radiation efficiency and S_{PV} is day lighting area. $\theta(t)$

is solar radiation intensity at time t.

(3) SC storage capacity model:

$$S_{sc}(t) = S_{sc}(t-1) + (g_{ch}(t)\eta_{ch} - \frac{g_{dis}(t)}{\eta_{dis}}) \tag{3}$$

In the formula, $S_{SC}(t)$ is storage capacity of the SC at time t. $g_{ch}(t)$ is SC charging power andη_{ch} is the charging efficiency of SC. $g_{dis}(t)$ is SC discharging power andη_{dis} is discharging efficiency of SC.

(4) Price-based DR (PSDR) model for electric loads

Electricity load PSDR is mainly through the introduction of peak-valley time-sharing tariff, forming a reasonable electricity consumption guidance for customers, so as to achieve peak and valley reduction of the load curve, price elasticity coefficients for different time periods based on peak and valley time-sharing tariffs to derive the post-response electric load.

$$e_{on} = \frac{\Delta Q_{on}(t)}{Q_{on}(t)} \frac{P_{on}(t)}{\Delta P_{on}(t)}, t \in T_{on} \tag{4}$$

$$e_{mind} = \frac{\Delta Q_{mind}(t)}{Q_{mind}(t)} \frac{P_{mind}(t)}{\Delta P_{mind}(t)}, t \in T_{mind} \tag{5}$$

$$e_{off} = \frac{\Delta Q_{off}(t)}{Q_{off}(t)} \frac{P_{off}(t)}{\Delta P_{off}(t)}, t \in T_{off} \tag{6}$$

In the formulas, e_{on}, e_{mind} and e_{off} are price elasticity coefficients for peak, flat and valley hours of electricity consumption. $\Delta Q_{on}(t)$, $\Delta Q_{mind}(t)$ and $\Delta Q_{off}(t)$ are the change amount of electrical load during peak, flat and valley hours. $Q_{on}(t)$, $Q_{mind}(t)$ and $Q_{off}(t)$ are original electrical load during peak, flat and valley hours. $P_{on}(t)$, $P_{mind}(t)$ and $P_{off}(t)$ are the original tariff for peak, flat and valley hours. $\Delta P_{on}(t)$, $\Delta P_{mind}(t)$ and $\Delta P_{off}(t)$ are the amount of electricity price change during peak, flat and valley hours.

3.2 Heating Equipment Output and Heat Load Incentive-Based DR Model

In this chapter, the heat output is mainly from the electric boiler and heat storage tank. When the electricity price is low, electric boilers can work with large power and surplus heat load can be stored in the thermal storage tank. When the electricity price is high, electric boilers can work with small power or stop working, with the heat storage tank releasing heating. With adjusting the indoor set temperature to reduce the heat load, VPPs provide economic compensation to responsive customers.

(1) EB power output model:

$$Q_{EB}(t) = g_{EB}(t) \cdot \eta_{EB} \tag{7}$$

In the formula, $Q_{EB}(t)$ is heating power of EB and $g_{EB}(t)$ is electricity consumed by the EB at time $t.\eta_{EB}$ is the electrical and thermal conversion efficiency.

(2) Heat storage capacity model for HS:

$$S_{HS}(t) = (1 - \eta_{HS})S_{HS}(t-1) + (Q_{in}(t)\eta_{in} - \frac{Q_{out}(t)}{\eta_{out}}) \tag{8}$$

In the formula, $S_{HS}(t)$ is the heat storage capacity of the HS at time t andη_{HS} is heat storage loss rate. $Q_{in}(t)$ is heat storage power at time t, and $Q_{out}(t)$ is exothermic power at time $t.\eta_{in}$ andη_{out} are the efficiency of heat storage and heat release of HS.

(3) Incentive-based DR (IBDR) model for heat loads:

The main factors affecting the heat load in winter are the heat dissipation caused by the difference between indoor and outdoor temperatures, the heat generated by the equipment and human body inside the building, and the heat dissipation caused by air infiltration. Therefore, the heat load response model is established as follows.

$$Q_{heart}(t) = \sum_{t=1}^{T} Q_{HT}(t) + Q_{INF}(t) - Q_{IH}(t) \tag{9}$$

where $Q_{heart}(t)$ is heat load required for the house; T is response time; $Q_{HT}(t)$ is heat conduction from the envelope at t time; $Q_{INF}(t)$ is heat consumption by air infiltration at t time; $Q_{IH}(t)$ is heat generation from indoor heat source at t time.

$$Q_{HT}(t) = \partial KA(T_{PMV} - T_{out}(t)) \tag{10}$$

$$Q_{INF}(t) = C_{air}\rho_{air}NSH(TPMV - T_{out}(t)) \tag{11}$$

$$Q_{IH} = Q_{ine} + Q_{inh} \tag{12}$$

In the formulas, ∂ is enclosure structure temperature difference correction factor. K is heat transfer coefficient of enclosure structure. A is enclosure area and T_{PMV} is room temperature. $T_{out}(t)$ is outdoor temperature at time t. C_{air} is specific heat capacity of air and ρ_{air} is air density. N is number of air changes and S is house size. H denotes house interior height and Q_{ine} is electrical equipment heat generation. Q_{inh} is human heat generation.

$$K = \frac{1}{\frac{1}{\partial_n} + \sum_{\partial_\lambda}^{\sigma} + R_k + \frac{1}{\partial_w}} \tag{13}$$

where ∂n is heat transfer coefficient of the inner surface of the envelope. σis material thickness of each layer of the envelope. ∂_λ indicates material thermal conductivity correction factor and λ is thermal conductivity of the material of each layer of the envelope. R_K denotes enclosed space layer thermal resistance and ∂w is heat transfer coefficient of the outer surface of the envelope.

Substituting (10)–(12) into (9), the model of heat load IBDR is formulated in (14).

$$Q_{\text{heart}}(t) = \sum_{t=1}^{T} (C_{\text{air}}\rho_{\text{air}}NSH + \partial KA)(T_{\text{PMV}} - T_{\text{out}}(t)) - Q_{\text{ine}} - Q_{\text{inh}} \tag{14}$$

3.3 Customer Comfort Modeling

While implementing DR, customer comfort should be considered, where previous studies mainly ignored the impact of customer comfort on the actual participation of DERs in DR decisions. Since the heat load can be changed by changing the indoor temperature to achieve the DR, considering that the perception of the ambient temperature has certain ambiguity, this chapter introduces the average thermal sensory predicted mean vote (PMV) to indicate the customer's satisfaction with the external ambient temperature. PMV index is an evaluation index to indicate the thermal response, representing the average of most customers' hot and cold sensations in the same environment, which is specifically divided into seven grades. PMV = −1, −2, −3 indicate that a customer feels cold, cool, slightly cool. PMV = 0, 1, 2, 3 indicate that a customer feels comfortable, slightly hot, hotter, hot, as shown in Fig. 2.

The relationship between indoor temperature and PMV index when other factors are at comfort level is formulated in (15).

$$\lambda_{\text{PMV}} = \begin{cases} 0.3895(T_{\text{PMV}} - 26)\,, T \geq 26 \\ 0.4065(-T_{\text{PMV}} + 26), T < 26 \end{cases} \tag{15}$$

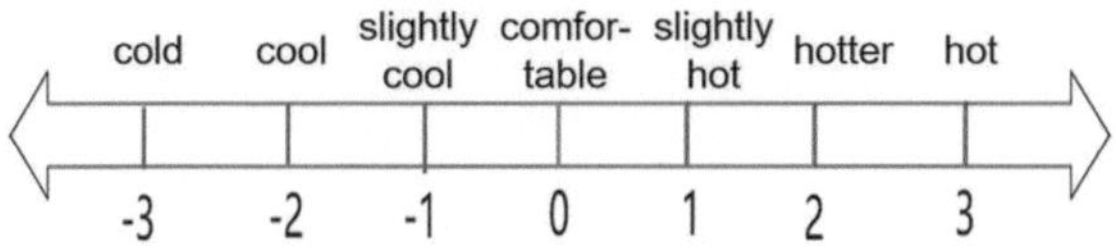

Fig. 2 The relationship between PMV index and user comfort

When the indoor temperature is 26 °C, λ_{PMV} is equal to 0, which means the customer's comfort is the highest at this time. The larger the λ_{PMV} is, the stronger the customer's heat feeling is, and the larger the capacity for implanting heat load DR.

According to the ISO-7730 standard, PMV index between −0.5 and 0.5 is suitable for human body. Therefore, corresponding to the (15), it can be obtained that corresponding indoor temperature is from 24.8 to 27.3 °C.

Since there are physiological differences between different customers and their sensations of heat and cold are not exactly the same, the predicted percentage of dissatisfied (PPD) index can be utilized to indicate the percentage of customers who are dissatisfied with the ambient temperature, where a quantitative relationship between PMV and PPD is formulated in (16).

$$PPD = 100 - 95 \times \exp[-(0.03353\lambda_{PMV}^4 + 0.2179\lambda_{PMV}^2)] \tag{16}$$

According to (16), when the PMV index is equal to 0, the customer comfort is the highest, at this time the lowest PPD value is 5%. That is to say, with the lowest dissatisfaction with the ambient temperature, the PMV index gradually approaches to −3 or 3, and customer comfort decreases where the PPD value gradually increases and the maximum value can be 100%. Customer comfort and customer dissatisfaction with ambient temperature directly affect the willingness of customers to participate in DR. The higher the customer comfort and the lower the dissatisfaction with ambient temperature, the higher the willingness of customers to participate in DR, and vice versa, as illustrated in Fig. 3.

In this chapter, for the sake of calculation simplicity, the customer willingness to participate in DR is set to meet the 0–1 probability distribution. When the ambient temperature is from 24.8 and 27.3 °C, customers will participate in DR. If the environment temperature is lower than 24.8 °C or higher than 27.3 °C, customers will not participate in DR. The specific formula is expressed as (17).

$$K = \begin{cases} 0, & T_{PMV} < 24.8\,°\mathrm{C} \cup T_{PMV} > 27.3\,°\mathrm{C} \\ 1, & 24.8\,°\mathrm{C} \le T_{PMV} \le 27.3\,°\mathrm{C} \end{cases} \tag{17}$$

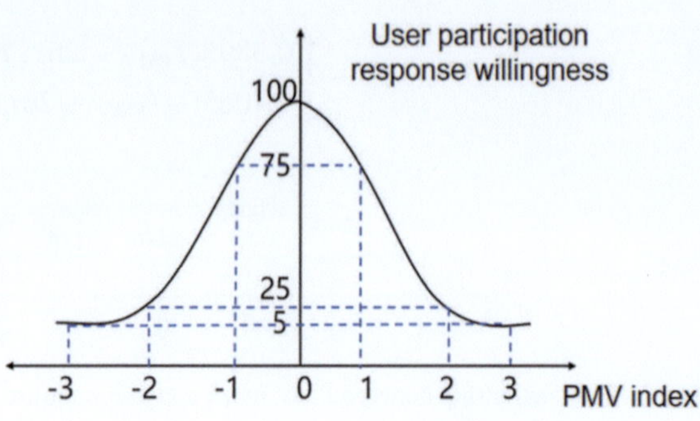

Fig. 3 The relationship between customers' willingness to participate in DR and the PMV index

In the formula, K denotes the probability of customer participation in DR. $K = 0$ means that a customer is not involved in DR while $K = 1$ indicates customer participation in DR.

4 Operation Optimization Model of VPPs

4.1 Objective Function

In this chapter, with the objective of minimizing the operating cost of VPPs, the objective function is established as (18).

$$C_{\mathrm{VPP}} = C_{\mathrm{WPP}}(t) + C_{\mathrm{PV}}(t) + C_{\mathrm{SC}}(t) + C_{\mathrm{HS}}(t) + C_{\mathrm{EB}}(t) + C_{\mathrm{M}}(t) + C_{\mathrm{IBDR}}(t) \quad (18)$$

In the formula, C_{VPP} is total cost of a VPP. $C_{\mathrm{WPP}}(t)$ is cost of wind power generation at time t. $C_{\mathrm{PV}}(t)$ denotes cost for PV power generation at time t. $C_{\mathrm{SC}}(t)$ is the cost of power output from the power storage unit at time t. $C_{\mathrm{HS}}(t)$ is cost of heat storage tank at time t. $C_{\mathrm{EB}}(t)$ is cost of electric boiler at time t. $C_{\mathrm{M}}(t)$ is cost of electricity purchased from the grid by a VPP at time t. $C_{\mathrm{IBDR}}(t)$ is heat load IBDR cost.

$$C_{\mathrm{WPP}}(t) = K_{\mathrm{WPP}} g_{\mathrm{WPP}}(t) \quad (19)$$

$$C_{\mathrm{PV}}(t) = K_{\mathrm{PV}} g_{\mathrm{PV}}(t) \quad (20)$$

$$C_{\mathrm{SC}}(t) = K_{\mathrm{SC}} |g_{\mathrm{SC}}(t)| \quad (21)$$

$$C_{\mathrm{HS}}(t) = K_{\mathrm{HS}} |Q_{\mathrm{HS}}(t)| \quad (22)$$

$$C_{\mathrm{EB}}(t) = K_{\mathrm{EB}} Q_{\mathrm{EB}} + P(t) g_{\mathrm{EB}}(t) \quad (23)$$

$$C_{\mathrm{M}} = P_{\mathrm{M}}(t) g_{\mathrm{M}}(t) \quad (24)$$

$$C_{\mathrm{IBDR}}(t) = \Delta Q_{\mathrm{heart}}(t) P_{\mathrm{heart}} \quad (25)$$

In these formulas, K is equipment operating cost factor. $P(t)$ is electricity price at time t and P_{M} is grid electricity sales prices. $g_{\mathrm{M}}(t)$ is electricity purchase of a VPP at time t and P_{heart} is price reduction of heat load compensation. $\Delta Q_{\mathrm{heart}}(t)$ presents heat load reduction.

4.2 Constraint Conditions

(1) Electrical and thermal power balance:

$$g(t) = g_{\text{WPP}}(t) + g_{\text{PV}}(t) + g_{\text{SOC}}(t) - g_{\text{EB}}(t) \tag{26}$$

$$Q_{\text{EB}}(t) - Q_{\text{heart}}(t) = Q_{\text{HS}}(t) \tag{27}$$

where $g(t)$ is total electric load demand.

(2) Wind power generators, photovoltaic generators, electric boiler operation constraints:

$$g_{\text{WPP}}^{\min} \leq g_{\text{WPP}}(t) \leq g_{\text{WPP}}^{\max} \tag{28}$$

$$g_{\text{PV}}^{\min} \leq g_{\text{PV}}(t) \leq g_{\text{PV}}^{\max} \tag{29}$$

$$Q_{\text{EB}}^{\min} \leq Q_{\text{EB}}(t) \leq Q_{\text{EB}}^{\max} \tag{30}$$

In these formulas, $g_{\text{WPP}}^{\min}$ and $g_{\text{WPP}}^{\max}$ are the lower power output limit and higher power output limit of a wind power plant. $g_{\text{PV}}^{\min}$ and $g_{\text{PV}}^{\max}$ are lower power output and higher power output limit of a photovoltaic equipment. $Q_{\text{EB}}^{\min}$ and $Q_{\text{EB}}^{\max}$ are lower power limit and higher power limit of an electric boiler.

(3) Electricity storage device, heat storage device operating constraints:

$$g_{\text{SC}}^{\min} \leq g_{\text{SC}}(t) \leq g_{\text{SC}}^{\max} \tag{31}$$

$$S_{\text{SC}}^{\min} \leq S_{\text{SC}}(t) \leq S_{\text{SC}}^{\max} \tag{32}$$

$$Q_{\text{HS}}^{\min} \leq Q_{\text{HS}}(t) \leq Q_{\text{HS}}^{\max} \tag{33}$$

$$S_{\text{HS}}^{\min} \leq S_{\text{HS}}(t) \leq S_{\text{HS}}^{\max} \tag{34}$$

where $g_{\text{SC}}^{\min}$ and $g_{\text{SC}}^{\max}$ are the minimum and maximum charging/discharging power of the battery. $g_{\text{SC}}(t)$ is the actual charging/discharging power of the battery at time t. $S_{\text{SC}}^{\min}$ and $S_{\text{SC}}^{\max}$ are the minimum and maximum battery storage capacity. $Q_{\text{HS}}^{\min}$ and $Q_{\text{HS}}^{\max}$ are the minimum and minimum power of heat storage tank. $Q_{\text{HS}}(t)$ is actual power of the heat storage tank at time t. $S_{\text{HS}}^{\min}$ and $S_{\text{HS}}^{\max}$ are the minimum and maximum heat storage capacity of heat storage tank.

(4) Customer comfort constraints:

When heating, to maintain the indoor temperature at the appropriate state for customers, the relationship between the heat generated by an electric boiler and the indoor temperature is formulated in (35)–(37).

$$\frac{\mathrm{d}T_{\mathrm{PMV}}(t)}{\mathrm{d}t}=\frac{Q_{\mathrm{EB}}(t)\Delta T_{\mathrm{PMV}}+R_{\mathrm{k}}T_{\mathrm{PMV}}(t)}{\rho_{\mathrm{air}}} \tag{35}$$

$$T_{\mathrm{PMV}}^{\min}\leq T_{\mathrm{PMV}}(t)\leq T_{\mathrm{PMV}}^{\max} \tag{36}$$

$$\lambda_{\mathrm{PMV}}^{\min}\leq \lambda_{\mathrm{PMV}}(t)\leq \lambda_{\mathrm{PMV}}^{\max} \tag{37}$$

In the formulas, $T_{\mathrm{PMV}}^{\min}$ and $T_{\mathrm{PMV}}^{\max}$ are the minimum and maximum indoor temperature. $\lambda_{\mathrm{PMV}}^{\min}$ and $\lambda_{\mathrm{PMV}}^{\max}$ are the minimum and maximum PMV indicator.

5 Case Studies

5.1 Basic Data

The parameters of wind turbine and photovoltaic equipment aggregated in a VPP are shown in Table 1. The parameters of energy storage equipment are shown in Table 2 and the parameters of electric boiler equipment are presented in Table 3. The purchase and sale price of electricity from the grid is shown in Table 4. Considering that the short-term temperature forecasting can basically meet the heat load forecasting needs, this chapter takes the day-ahead weather forecasting information as the benchmark when calculating the heat load demand and ignores the influence of relevant uncertainties. The initial tariff is unified at 0.4 yuan/kWh, and the price elasticity coefficients adopted for the peak-to-valley tariff are shown in Table 5. Peak tariff periods are 11:00–15:00, 19:00–21:00, and the tariff is 0.8 yuan/kWh. The flat tariff periods are 08:00–10:00, 16:00–18:00, 22:00–23:00, and the tariff is 0.5 yuan/kWh. The valley tariff periods are 24:00–07:00, and the tariff is 0.2 yuan/kWh. The structural parameters of the house [18] are shown in Table 6.

Table 1 The parameters of the wind turbine and photovoltaic equipment

WPP		PV	
Power limit	210 MW	Power limit	150 MW
Lower power limit	0	Lower power limit	0
Maintenance factor	45 yuan/MW	Maintenance factor	9.6 yuan/MW

Table 2 The parameters of energy storage equipment

SC		HS	
Rated capacity	100 MWh	Rated capacity	100 MWh
Maximum storage capacity	100 MWh	Maximum storage capacity	100 MWh
Minimum energy storage	0 MWh	Minimum energy storage	0 MWh
Charging efficiency	0.95	Thermal storage efficiency	0.88
Discharge efficiency	0.95	Exothermic efficiency	0.88
Maximum charging power	100 MW	Maximum heat storage power	100 MW
Maximum discharge power	100 MW	Maximum exothermic power	100 MW
Maintenance factor	20 yuan/MW	Maintenance factor	20 yuan/MW

Table 3 The electric boiler equipment parameters

EB	
Power limit	200 MW
Lower power limit	0
Electric rotary heat efficiency	0.9
Maintenance factor	10 yuan/MW

Table 4 The electricity purchase and sale prices of power grid

Time period	Grid electricity sales price (yuan/MWh)	Electricity network purchase electricity price (yuan/MWh)
Peak hours	990	800
Usual period	530	400
Valley hours	170	100

Table 5 The price elasticity coefficient

Time period	Price elasticity coefficient
Peak hours	−0.25
Usual period	−0.1
Valley hours	0.08

5.2 *Operation Scenario Analysis of VPPs*

In order to verify the effectiveness of the operating cost of VPPs through integrated electricity and heat DR as proposed in this chapter, the following three scenarios are considered.

Scenario 1: no DR measures are considered. The original operation of the VPP is used as a benchmark to compare with the operation of the VPP in Scenarios 2 and 3.

Table 6 The structural parameters of buildings

Parameters	Numerical value
House size S/m^2	100
House height H/m	3
Indoor air volume V/m^3	300
Wall thickness δ/m	0.37
Enclosure area A/m^2	290
Enclosure structure temperature difference correction factor α	1.0
Correction coefficient of thermal conductivity of wall materials α_λ	1.2
Heat transfer coefficient of the inner surface of the envelope α_n(W/m^2 k)	8.7
Heat transfer coefficient of the outer surface of the envelope α_w(W/m^2 k)	23
Enclosed air interlayer thermal resistance R_k(m^2 k/W)	0
Thermal conductivity of wall materials λ (W/m k)	0.81
Air density ρ_{air} (kg/m^3)	1.4
Specific heat capacity of air Cair (KJ/(kg °C))	1.0
Number of air changes N (times/h)	0.5
Electrical equipment heat generation Q_{ine}/kW	0.2
Human heat generation Q_{inh}/kW	0.05
Comfortable state human skin temperature T_S/°C	33.5
Human energy metabolic rate M/(W/m^2)	80
Garment thermal resistance I_{cl}/((m^2 °C)/W)	0.11

Scenario 2: only electric load PBDR is considered. Based on the original operation of the VPP, the PBDR of the electrical load is implemented to investigate a single type of DR on the operation of the VPP.

Scenario 3: both of the electric load PBDR and thermal load IBDR are considered. Multiple types of DR considering customer comfort are implemented to investigate the impact of integrated multi-load DRs on the operation of the VPP.

The electric load of the VPP in Scenario 1 is mainly supplied by wind power generation and solar power generation, where the forecasted power generation curve is presented in Fig. 4. The heat load is mainly supplied by the electric boiler, where the forecasted power curve is illustrated in Fig. 5. The forecasted power of energy storage device and thermal storage tank is shown in Fig. 6, where a negative number indicates that the energy storage device is in charging state. The VPP transactions with the grid are shown in Fig. 7, where a negative number indicates that the VPP sells surplus electricity to the power grid.

The VPP control center in Scenario 2 implements the PBDR for the electric load, where the electricity changes according to the price elasticity coefficients as presented in Fig. 8.

After implanting the electric load DR, the electric load demand during the valley tariff and flat tariff period increased by 407.9 MW compared with Scenario 1, while

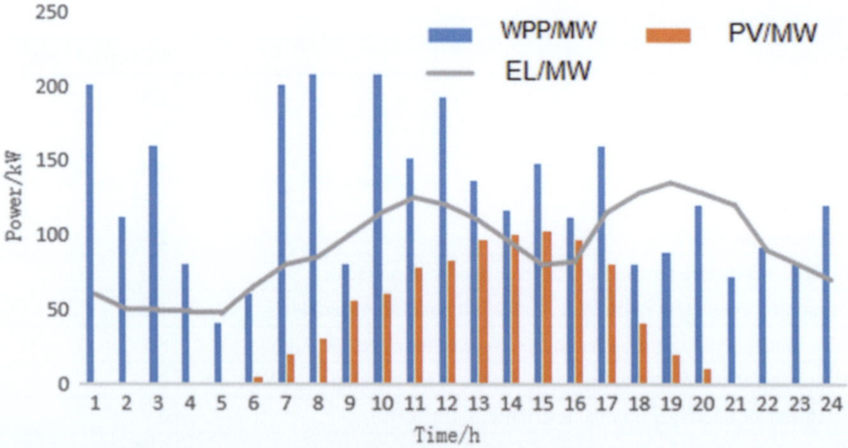

Fig. 4 The output of wind power and solar power with electric load

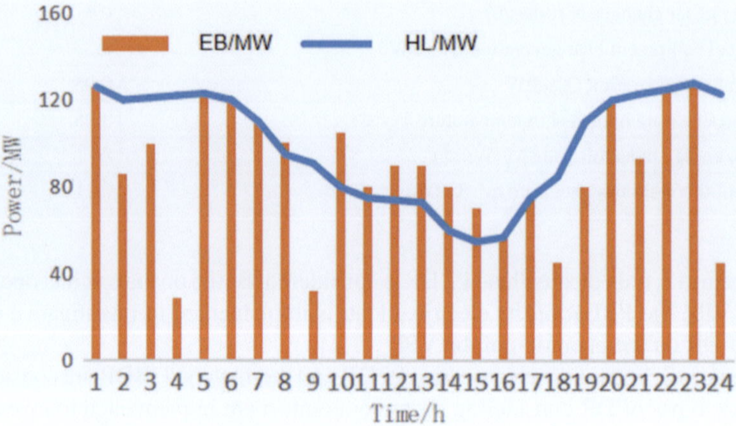

Fig. 5 The output of the electric boiler and the heat load

the electric load demand during the peak tariff period decreased by 457.3 MW compared with Scenario 1. The peak-to-valley ratio of electric load decreased from 2.87 before the response to 2.15, thus realizing the peak and valley reduction of the electric load, and relieving the operating pressure of the VPP. At the same time, due to the peak-shaving of the electric load, part of the saved electricity is used to improve the heat production of the electric boiler, whose output is increased by 147 MW compared with Scenario 1. The total power of the energy storage device remains unchanged. The power follows the changes of the electric load and the power of the electric boiler with slight changes, as shown in Fig. 9. This in turn affects the volume of transactions between the VPP and the power grid, where the electricity

Fig. 6 The power of energy storage device and thermal storage tank

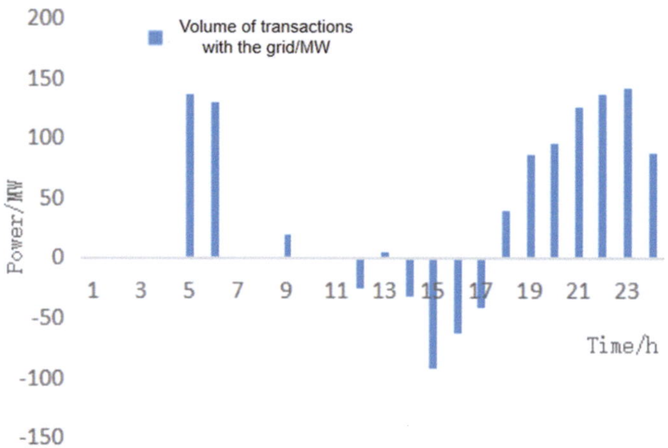

Fig. 7 The VPP transactions with the grid

purchased by the VPP increases by 12.34 MW, mostly occurring in the valley hours. The electricity sales of the VPP increases by 54.13 MW, as presented in Fig. 10.

Scenario 3 implements PBDR for electric loads along with appropriate IBDR for thermal loads, considering the lowest possible compensation costs while reducing the operating costs of the VPP. Therefore, the common peak period of electric and heat load demand, i.e., 19:00 to 23:00, is selected for IBDR. However, under different PMV indicators, the indoor set temperature is different with distinct required heat load. The relationship between customer comfort and heat load demand is shown in Fig. 11.

It is obvious from Fig. 11 that under the premise of ensuring customer comfort, the smaller heat load demand occurs when λ_{PMV} is euqal to -0.5 with the smallest pressure of heating equipment. Since the heat load demand is linearly related to the

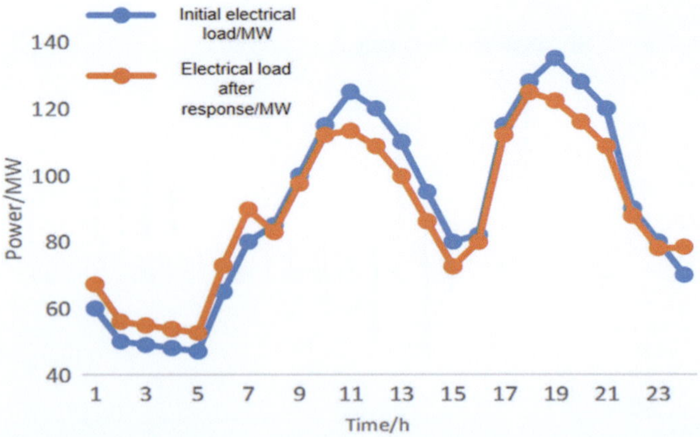

Fig. 8 The electric load change curve

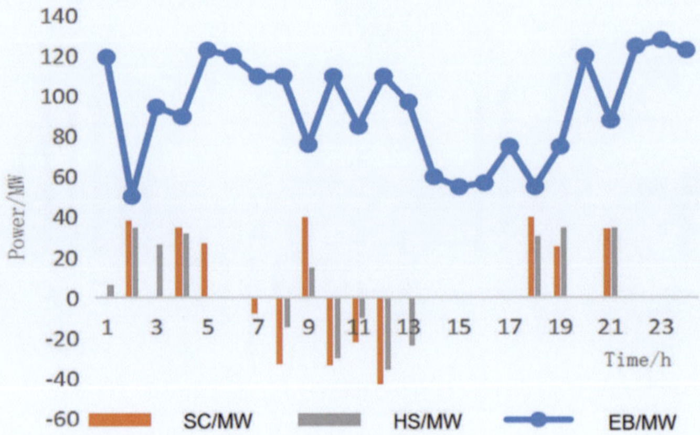

Fig. 9 The power of the electric boiler and energy storage

cost of electric boiler, the cost of electric boiler is the smallest at this time. The heat load demand is reduced by 50 MW during this time period, and the electric boiler output is reduced by 50 MW compared with Scenario 2, thus reducing the transaction volume between the VPP and the power grid. The electricity purchase is reduced by 53.89 MW compared with Scenario 2, as shown in Fig. 12, which further reduces the operating cost of the VPP and realizes the goal of optimal operation of the VPP.

Based on the data of each equipment operation cost parameter, grid purchase and sale price, and IBDR compensation cost, the operation costs of the VPP under three scenarios are presented in Table 7.

The specific operating costs of the VPP under different scenarios are analyzed as follows. The operating costs of the wind power generation and solar generation are 13.5 and 0.83 million yuan in the three scenarios, respectively. Since the wind

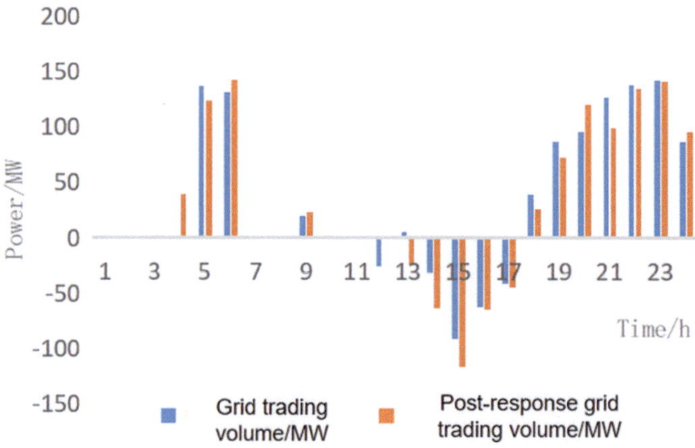

Fig. 10 Comparison of transaction volume between the VPP and power grid

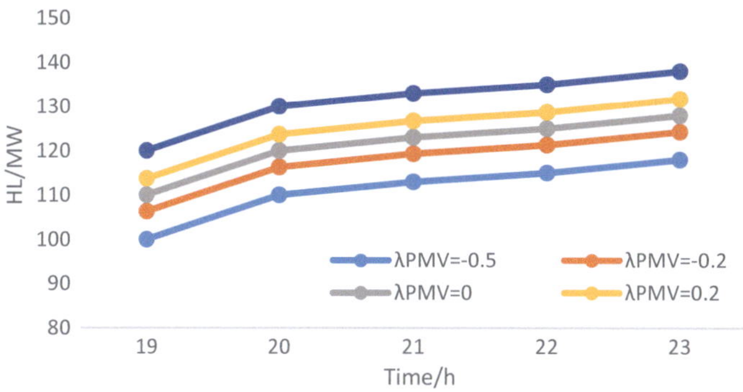

Fig. 11 The heat load demand under different customer comfort

power generation and solar generation provide the electrical load for the VPP in the three scenarios, they are in the same external environment and have the same equipment parameters, so the operating costs are the same in the three scenarios. Scenarios 2 and 3 both consider the PBDR of electric load, where the electric load achieves peak reduction and valley filling, thus affecting the charging power of the energy storage device. The operating costs of the energy storage device and heat storage tank show an overall upward trend with the change of the electric load, rising from 0.2 to 0.76 million yuan and 0.66 million yuan, respectively. The electric load demand changes in Scenario 2 while the heat load demand remains unchanged directly affecting the heat output of the electric boiler. The IBDR of the heat load considering customer comfort in Scenario 3 further reduces the heat load demand and thus the power of the electric boiler changes where the overall operating cost is reduced from 2.3 to 2.2 million yuan. Since the IBDR is to give financial compensation to responding customers, the DR cost in Scenario 3 is 0.0026 million yuan. The

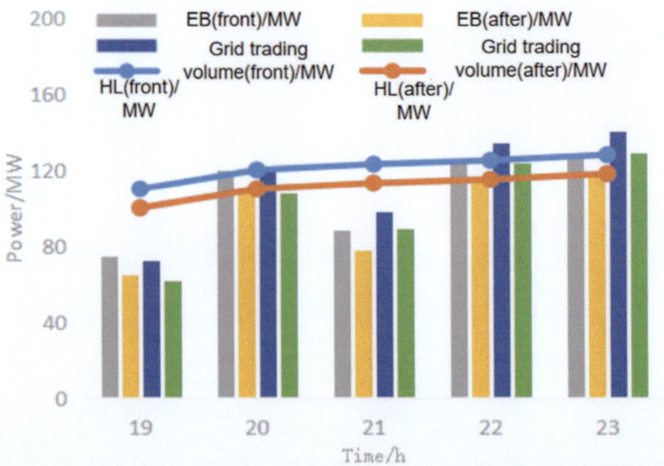

Fig. 12 The output of electric boiler and energy storage device with trading volume with power grid

Table 7 Comparison of operation costs of the VPP under three scenarios

Cost/million yuan	Scenario 1	Scenario 2	Scenario 3
$C_{WPP}(t)$	13.5	13.5	13.5
$C_{PV}(t)$	0.83	0.83	0.83
$C_{SC}(t)$	0.2	0.76	0.76
$C_{HS}(t)$	0.2	0.66	0.66
$C_{EB}(t)$	2.1	2.3	2.2
$C_{IBDR}(t)$	0	0	0.0026
$C_{M}(t)$	67.4	33.6	27.6
C_{VPP}	84.23	51.65	45.55

reduction of electric load demand in Scenario 2 reduces the VPP's transaction with the power grid. Scenario 3 reduces the heat load demand on the basis of Scenario 2, thus further reducing the VPP's transaction with the power grid. Therefore, the VPP's cost of purchasing electricity from the power grid is reduced from the initial 67. to 33.6 million yuan and finally to 27.6 million yuan. By comparing the total cost of the VPP, it can be found that the total operating cost of the VPP in Scenario 3 is the lowest, i.e., the implementation of DR measures for electrical and thermal loads effectively achieves load shifting and reduction. With optimizing the operating costs of the VPPs, the proposed model provide a reference for optimal operation of VPPs considering customer comfort.

6 Conclusion

In this chapter, based on the peak-valley time-sharing tariff and the predicted mean vote index, the integrated DR with PBDR and IBDR of the electric and thermal loads in a VPP are implemented respectively to realize the reducing operating cost of a VPP. Moreover, VPPs can purchase electricity or sell electricity with the power grid, which can help consume abandoned wind power and solar power. Furthermore, it can help increase the revenue of VPPs and provide a reference for VPPs to optimize the scheduling problem.

The proposed model considers customer comfort while implementing integrated DR in VPPs, which can improve the motivation of customers to make response, so that the actual response is close to the expected response to achieve peak reduction and valley filling of the load and the goal of optimizing the operating costs of VPPs.

References

1. A. Mnatsakanyan, S.W. Kennedy, A novel demand response model with an application for a virtual power plant. IEEE Trans. Smart Grid **6**(1), 230–237 (2015)
2. A. Thavlov, H.W. Bindner, Utilization of flexible demand in a virtual power plant set-Up. IEEE Trans. Smart Grid **6**(2), 640–647 (2015)
3. Z.M. Liang, Q. Alsafasfeh, T. Jin et al., Risk-constrained optimal energy management for virtual power plants considering correlated demand response. IEEE Trans. Smart Grid **10**(2), 1577–1587 (2019)
4. M. Vahedipour-Dahraie, H. Rashidizadeh-Kermani, M. Shafie-Khah, J.P.S. Catalão, Risk-averse optimal energy and reserve scheduling for virtual power plants incorporating demand response programs. IEEE Trans. Smart Grid **12**(2), 1405–1415 (2021)
5. Z. Azimi, R.-A. Hooshmand, S. Soleymani, Optimal integration of demand response programs and electric vehicles in coordinated energy management of industrial virtual power plants. Energy Storage **41**, 1–15 (2021)
6. L.W. Ju, Z.F. Tan, J.Y. Yuan et al., A bi-level stochastic scheduling optimization model for a virtual power plant connected to a wind-photovoltaic-energy storage system considering the uncertainty and demand response. Appl. Energy **171**(1), 184–199 (2016)
7. Z. Luo, S.H. Hong, Y.M. Ding et al., A data mining-driven incentive-based demand response scheme for a virtual power plant. Appl. Energy **239**(1), 549–559 (2019)
8. S. Aras, From demand response in smart grid toward integrated demand response in smart energy hub. IEEE Trans. Smart Grid **7**(2), 650–658 (2016)
9. P. Samaneh, H. Mahmoud-Reza, Optimal planning and scheduling of energy hub in presence of wind, storage and demand response under uncertainty. Int. J. Electr. Power Energy Syst. **80**, 219–239 (2016)
10. H. Yang, M. Li, Z. Jiang, P. Zhang, Multi-time scale optimal scheduling of regional integrated energy systems considering integrated demand response. IEEE Access **8**, 5080–5090 (2020)
11. Z. Zhang, C. Wang, H. Lv, F. Liu, H. Sheng, M. Yang, Day-ahead optimal dispatch for integrated energy system considering power-to-gas and dynamic pipeline networks. IEEE Trans. Ind. Appl. **57**(4), 3317–3328 (2021)
12. G. Zhang, Z. Shen, Z. Li, L. Wang, Energy scheduling for networked microgrids with co-generation and energy storage. IEEE Internet Things J. **6**(5), 7722–7736 (2019)
13. W. Fan, L. Huang, Z. Tan, F. Xue, G. De, X. Song, B. Cong, Multi-objective optimal model of rural multi-energy complementary system with biogas cogeneration and electric vehicle considering carbon emission and satisfaction. Sustain. Cities Soc. **74**, 1–10 (2021)

14. Q. Yang, P. Liu, J. Zhang, N. Dong, Combined heat and power economic dispatch using an adaptive cuckoo search with differential evolution mutation. Appl. Energy **307**, 1–12 (2022)
15. M.F. Tahir, C. Haoyong, K. Mehmood, N. Ali, J.A. Bhutto, Integrated energy system modeling of China for 2020 by incorporating demand response, heat pump and thermal storage. IEEE Access **7**, 40095–40108 (2019)
16. C. Shao, Y. Ding, P. Siano, Z. Lin, A framework for incorporating demand response of smart buildings into the integrated heat and electricity energy system. IEEE Trans. Industr. Electron. **66**(2), 1465–1475 (2019)
17. W. Tang, H. Yang, Optimal operation and bidding strategy of a virtual power plant integrated with energy storage systems and elasticity demand response. IEEE Access **7**, 79798–79809 (2019)
18. Z. Liang, Q. Alsafasfeh, T. Jin, H. Pourbabak et al., Risk-constrained optimal energy management for virtual power plants considering correlated demand response. IEEE Trans. Smart Grid **10**(2), 1577–1587 (2019)
19. D.T. Nguyen, L.B. Le, Joint optimization of electric vehicle and home energy scheduling considering user comfort preference. IEEE Trans. Smart Grid **5**(1), 188–199 (2014)
20. H.T. Dinh, D. Kim, An optimal energy-saving home energy management supporting user comfort and electricity selling with different prices. IEEE Access **9**, 9235–9249 (2021)
21. B. Zhou, Y. Zhang, T. Zang, W. Hua, Optimal dispatching of distribution network considering system flexibility and user thermal comfort. IEEE Access **9**, 107895–107908 (2021)
22. O. Erdinç, A. Taşcıkaraoğlu, N.G. Paterakis, Y. Eren, J.P.S. Catalão, End-user comfort oriented day-ahead planning for responsive residential HVAC demand aggregation considering weather forecasts. IEEE Trans. Smart Grid **8**(8), 362–372 (2017)

Chapter 6
Optimal Operation of Virtual Power Plants Participating in Auxiliary Service Market Coordinating with Energy Storage Allocation

With the rapid development of DERs, demand-side entities with high-power load aggregation, are actively acquiring opportunities to participate in electricity market transactions. VPPs with high regulating potential are typical forms of participation in the auxiliary service market. This chapter analyzes the bidding strategies and dispatching schemes of VPPs in the auxiliary service market to improve the capacity of VPPs, realize economic dispatching, and promote the resource utilization of VPPs. Specifically speaking, according to a series of rules for market entities to participate in the auxiliary service market transactions, the market bidding scheme of VPPs (including energy storage systems coordinated VPPs) is put forward. Then, the the conditional value at risk (CVaR) method is used to determine the coordinated energy storage capacity and efficiency of VPPs in the short term. Moreover, the transaction revenue, spread revenue, investment cost and opportunity cost are calculated according to the aggregation capacity, response situation and transaction cost. Finally, case studies are attached to this chapter, and the classic scenario of aggregated electric vehicles to form VPPs is taken as an example. The case studies suggest that the VPPs provide valley filling service by participating in the auxiliary service market, proving that the above strategies can improve the responsiveness of VPPs and increase the revenue.

1 Introduction

The global green-oriented transition of energy process is accelerating. All countries are vigorously developing the power market to promote the interconnection between the power supply side and the demand side, helping the power supply side to accommodate electricity form new energies, and alleviating power shortage on the demand side. Getting more third-party demand-side entities to access the auxiliary service market will enrich service types, improve service effect and promote active market transactions. As important players in the auxiliary service market, VPPs are

H. Jia et al., *Business Models and Reliable Operation of Virtual Power Plants*,
https://doi.org/10.1007/978-981-19-7846-3_6

of great significance for the diluted service costs, maintaining the stability of the power system, and even facilitating the development of renewable energies [1–3].

For demand-side entities' participation in the auxiliary service market, extensive research has focused on key technical breakthroughs at the market organization and dispatch. The auxiliary service market is becoming more and more open, so the VPPs formed by the aggregated DERs can well make market trading and provide practical services through the dispatching scheme [4]. Co-optimization of multiple markets, power systems, and the cyber network can maximize VPPs revenue [5]. Several studies have also shown that multiple VPPs obtain the optimal decision through the bi-level collaborative operation, which includes the bidding strategies and the economic dispatching instructions [6]. However, the above research is not comprehensive enough. While considering the bidding strategy and benefits, it is also necessary to consider the stochastic risk of aggregated DERs, which may cause some losses.

Moreover, due to strong regulation capacity of energy storage resources, energy storage resources have become an important means of power regulation. Many studies have also explored the coordination of energy storage resources and other resources, as well as the optimal scheduling problems. A bi-level hierarchical model [7] was built for the aggregation of DERs and energy storage to study participation in demand response. In [8], an accurate energy storage scheduling in real-time balancing stage was proposed based on a data-driven probabilistic characterization. Using a bi-level model can also optimize the location and size for energy storage systems participating in different markets [9].

The dynamic clustering algorithm can study the dynamic characteristics of VPPs integrated energy storage systems [10]. VPPs with energy storage systems participating in elasticity demand response optimization operation and bidding strategy was proposed in [11]. Large-scale VPPs aggregate the energy storage, and implement a day-ahead optimized scheduling [12] for electricity and heat. Comprehensive studies have found that energy storage and other distributed resources are integrated into VPPs [13], but the stability of power systems can be maintained through a centralized energy storage system [14]. This chapter will study the collaborative operation of centralized energy storage systems and VPPs to participate in the ancillary service market to further reduce deviations caused by VPPs and stochastic load characteristics.

This chapter first proposes a bidding strategy for VPPs to coordinate centralized energy storage and participate in the ancillary services market. Secondly, the short-term response capacity of VPPs and energy storage coordination capacity requirements is calculated based on method of least squares support vector machines (LSSVM) and method of the conditional value at risk (CVaR), respectively. Then, the benefits of VPPs collaborating centralized energy storage systems participating in the auxiliary service market and its calculation process are introduced. At last, a typical VPP aggregated with a large number of EVs is used as case study to analyze the corresponding response capacity and optimal operation with centralized energy storage systems.

2 Market Mechanism for VPPs Participating in Auxiliary Service Market

The VPPs and the power grid are connected through cyber networks and physical networks for power transmission and distribution, and the interaction is better reflected when aggregating DERs such as EVs. It is worth noting that when the centralized energy storage systems and VPPs independently participate in the interaction, they provide services with different performance at different costs. Centralized energy storage can directly provide auxiliary services with accurate response, with service costs being high. VPPs aggregate different distributed subjects, resulting in uncertainties that make the service process fluctuate, but the corresponding cost is relatively low.

2.1 Characteristic Analysis of DERs Aggregated in VPPs

A variety of third-party entities can provide auxiliary services with different performance. Through diversified control methods, third-party entities can conduct peak-shaving and valley-filling with different response speeds and capacities. The input construction, scheduling costs and benefit sources are different as well as the load fluctuations. Therefore, it is necessary to correctly classify and summarize different types of resources, which is the key to effective coordinated operation. Taking EVs as a type of resources aggregated in VPPs, Table 1 shows the feature comparison of EVs and centralized energy storage.

The operational reserves of VPPs themselves are of great uncertainties, which make VPPs not fully complete the task for maintaining the reliability of power systems. However, it is undeniable that VPPs are appropriate flexible demand-side

Table 1 Comparison of EVs and centralized energy storage

Items	EVs	Centralized energy storage (lithium-ion batteries)
Investment cost	185–333$/kW	1085–4100$/kW, 900–6200$/kWh
Benefit source	Load adjustment	Charging/discharge adjustment
Control method	User active response	Direct intelligent control
Response speed	Hourly and daily response	Minute and hourly response
Response speed	Large uncertainties	Accurate control
Dispatchability	A lot of constraints, Customers' requirements	Relatively few constraints
Application scenarios	Peak regulation, renewable energy consumption	Various types of ancillary services

resources, which can collect a variety of third-party entities to continuously and stably provide partly capacities for power systems, while centralized energy storage systems serve as suitable supplement, taking into account maintaing system reliability and economical operation.

The response capacity of VPPs can be accurately estimated through big data technologies, and the bidding strategy of VPPs aggregated with DERs and energy storage systems to participate in the auxiliary service market is proposed. Meanwhile, considering the changes of VPPs' capacity, the capacity requirements of the corresponding energy storage are adjusted, and different bidding strategies are designed to adapt to different aggregation scenarios. Moreover, the actual response to participate in the auxiliary service market also change with the bidding strategy and is mentioned below.

2.2 *Bidding Strategies for VPPs Participating in the Auxiliary Service Market*

VPPs' participation in the auxiliary service market needs to report the service time and expected response capacity to the trading center in advance as agreed. Take the valley filling service as an example, after the completion of the service, the actual response should be assessed. If the service failure is determined to a certain extent, the penalty will be accepted, and the service fee will be deducted. If it exceeds the tolerable range, it will be regarded as failing to provide the service and losing all the income. To avoid the failures that cannot be handled in time in the response process, it can be compensated by dispatching the capacity from the energy storage system, where the dispatching energy storage capacity needs to be determined. The calculation process is shown in Fig. 1.

VPPs with energy storage systems need to obtain the daily coordinated demand capacity of energy storage capacity, rather than the fuzzy number under long-term estimates. Such short-term demand capacity of energy storage can be learned through the conditional value at risk (CVaR). According to the above description, since the capacity curve that needs to be reported in advance, energy storage is a favorable supplement to VPPs.

Energy storage systems work with VPPs to participate in the auxiliary service market, and the configuration of energy storage systems also affects responsiveness. This section defines the concepts of two kinds of energy storage devices, one is promised energy storage, and the other is benefit energy storage. Promised energy storage is the coordinated energy storage capacity required to keep the load error within the market threshold while benefit energy storage capacity is the coordinated energy storage capacity required to make the response error of VPPs close to zero.

Not only to meet the needs of long-term energy storage applications, but also to consider the response of VPPs and the coordinated energy storage capacity in the short term, the following three bidding strategies are proposed. That is, the size

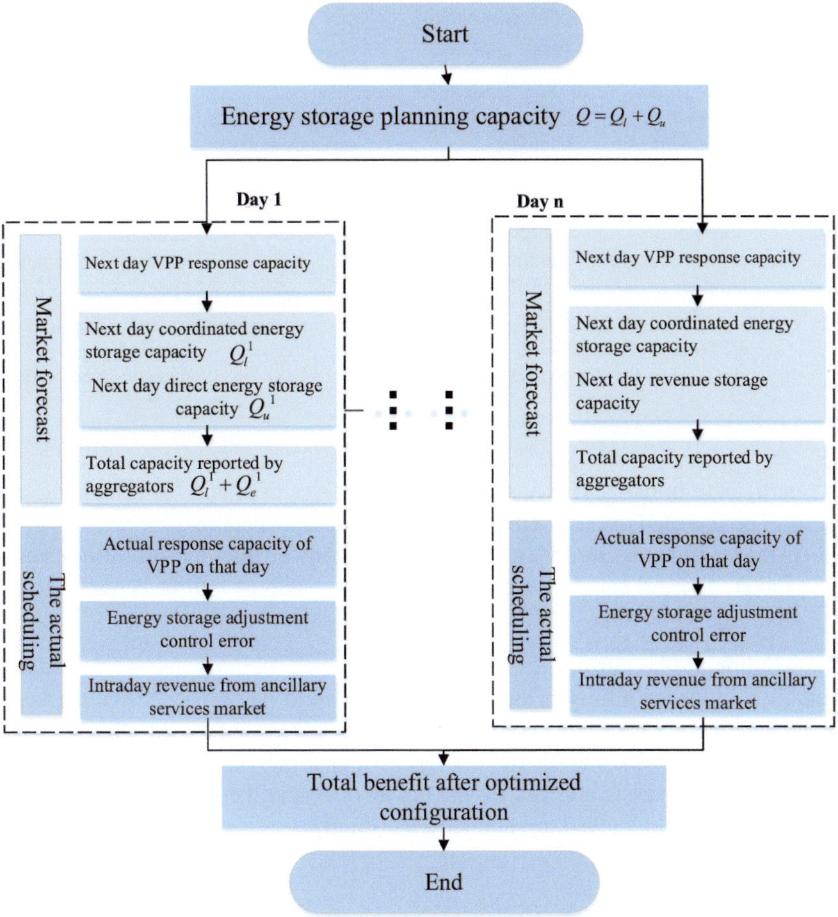

Fig. 1 The market strategy of energy storage coordinating with VPPs to participating in auxiliary service market

relationship of the configured energy storage capacity, the promised energy storage demand capacity and the benefit energy storage demand capacity comparison, to determine the plan on the day of the response.

(1) If the configured energy storage capacity is larger than the benefit energy storage demand capacity on that day, regulate the benefit energy storage capacity resources to coordinate VPPs participating in auxiliary service market transactions, and the remaining part is reported to the auxiliary service market independently.
(2) If the configured energy storage capacity is lower than the benefit energy storage capacity and higher than the promised energy storage demand capacity, the promised energy storage capacity will be regulated to coordinate the VPPs to

Table 2 Market response under different bidding strategies

Bidding strategy	Market response capacity of VPPs	Direct energy storage capacity	Price	Settlement gains and losses
(1)	VPPs + settlement gains and losses	Energy storage planning capacity—settlement gains and losses	Auxiliary services market clearing prices	Standard revenue
(2)	VPPs + promised energy storage	Energy storage planning capacity—promised energy storage		Standard revenue—settlement losses
(3)	VPPs	Energy storage planning capacity		Standard revenue—settlement losses

participate in the auxiliary service market transaction, and the remaining part will be reported to the auxiliary service market independently.

(3) If the configured energy storage capacity is lower than the promised energy storage demand capacity, all energy storage resources will be reported to the auxiliary service market directly and independently.

According to the current development of various auxiliary service markets in China, the influence ability of demand-side entities to participate in the market is small, with low market share. The quotation does not have reference value, but such demand-side entities have the right to priority clearing. Therefore, the source of income after delivery is affected by the bid amount and the settlement loss. Three bidding strategies in Table 2 are defined accordingly.

3 Measurement of Response Capacity

With the development of smart grid, VPPs integrate a wide variety of DERs to participate in the operation of power systems in the form of virtual plants, helping the overall system to achieve a wider range of functions and maintaining the stability of the power system. For example, the charging pile of EVs aggregation forms an important part of the VPPs, with the interaction between EVs and the grid, making the charging and discharging services connect to the grid through VPPs. This service pattern has been tested and popularized in multiple regions. This section takes a typical VPP as an example that aggregates a large number of EVs to measure corresponding response capacity.

3.1 General Situation of a VPP Aggregated with EVs

The response amount of EVs in the process of providing auxiliary services is determined by the overall scale and charging demand of EVs in the platform area, which is also affected by the direct response capacity of the energy storage system in VPPs. The amount of the response scale is measured from the short-term (daily) and long-term (annual) dimensions.

Big data technologies can accurately predict the short-term response ability of VPPs containing EVs, while the long-term response ability is predicted by historical short-term actual response data using statistical methods to simulate distribution functions.

3.2 Response Capacity of a VPP Aggregated with EVs

(1) Short-term response capacity of a VPP aggregated with EVs

The load balance of the power system requires peak shifting and valley filling services every day, and the participation of the demand-side entities is very important. For example, the EVs mentioned in this section provide valley filling when the auxiliary service market is opened. By predicting the load of EV aggregations, obtaining the daily load response capacity, and selecting the response capacity of energy storage according to the results to make up for the lack of response of EVs, this measure will improve the overall service quality of VPPs.

Least Square Support Vector Machines (LSSVM) is applicable to load prediction [15]. In this section, LSSVM will be used to perform intra-day load prediction and scenario analysis. Then, the overall deviation distribution analysis of VPPs providing auxiliary services is conducted. The coordinated response capacity under conditional risk is estimated from the statistical sample of the overall deviation distribution. The main process of LSSVM are highlighted below.

LSSVM is based on the principle of structural risk minimization to find the optimal w, b, and the optimization problem is as Eq. (1).

$$\min \frac{1}{2} w^T w + r \sum_{i=1}^{n} \xi_i^2 \tag{1}$$

where $r > 0$ is the penalty parameter and ξ_i is the relaxation variable.

Constraints are formulated in Eq. (2).

$$y[w_T \cdot \varphi(x_i) + b] = 1 - \xi_i, \quad i = 1, 2, \ldots, n \tag{2}$$

Applying the Lagrange functions to solve the optimization problem:

$$L = \frac{1}{2}w^T \cdot w + r \cdot \frac{1}{2}\sum_{i=1}^{n} \xi_i^2 - \sum_{i=1}^{n} \alpha_i \{y_i[w^T \cdot \varphi(x_i) + b] - 1 + \xi_i\}$$

$$s.t. \quad \begin{cases} w = \sum_{i=1}^{n} \alpha_i y_i \varphi(x_i) \\ \sum_{i=1}^{n} \alpha_i y_i = 0 \\ \alpha_i = r\xi_i \\ y_i[w^T \cdot \varphi(x_i) + b] - 1 + \xi_i = 0 \end{cases} \tag{3}$$

where α_i is the Lagrange multiplier vector, and $\alpha_i > 0, i = 1, 2, \ldots, n$. The final prediction function can be obtained as shown in Eq. (4).

$$f(x) = \sum_{i=1}^{n} \alpha_i K(x, x_i) + b \tag{4}$$

where $K(x_i, x_j) = \varphi(x_i)^T \varphi(x_j)$ is the kernel function satisfying the Mercer condition.

(2) Response capacity of a VPP aggregated with EVs

Long-term response ability is the cumulative comprehensive result of short-term response effect within a certain period of time. The response capacity distribution has been proved to be subject to the normal distribution function. By fitting the normal distribution function, the distribution function of the daily charging capacity of EVs can be obtained as shown in Eq. (5).

$$F(q) = \frac{1}{\sqrt{2\pi}\sigma} \int_0^q \exp\left(-\frac{(q-\mu)^2}{2\sigma^2}\right) \tag{5}$$

where q is the size of short-term response capacity. μ and σ are the mean and standard deviation parameters of normal distribution.

Further simulations using the distribution function yield multiple short-term response capacity samples, which can be summed to study the long-term response capacity of a VPP aggregated with EVs.

3.3 Energy Storage Response Capacity

CVaR is an advanced method on the value-at-risk (VaR) method proposed by Rockafeller, Uryasev and other scholars. The independent assumption and simple calculation process of normal distribution function make CVaR widely adopted. The basic mathematical expression is as Eq. (6).

$$CVaR_{\beta} = E\left[f|f \geq VaR_{\beta}\right] \tag{6}$$

where f represents portfolio losses and β represents confidence level.

There is an inevitable gap between the expected response amount and the actual response amount of VPPs, which is also the key point of assessment and even directly affects the income. The capacity difference between expected and actual values is given by Eq. (7) and corrected by CVaR to the more applicable Eqs. (8)–(10). Therefore, at a certain level of confidence, the response capacity of VPPs is supplemented by energy storage systems for improving the overall service effect and ensuring the revenue.

$$f(t, P) = P_t - E_t \tag{7}$$

$$P(f(t, P) \leq VaR(p)) = 1 - p \tag{8}$$

$$\tilde{F}_{\beta}(t, P) = a + \frac{1}{m(1-\beta)} \sum_{k=1}^{m} \left[f\left(t, P^k\right) - a\right]^+ \tag{9}$$

$$[f(t, P) - a]^+ = \begin{cases} f(t, P) - a, & f(t, P) > a \\ 0, & f(t, P) \leq a \end{cases} \tag{10}$$

where $f(t, P)$ is the gap power; P_t is the response power value to be provided at time t; E_t is the peak modulation power of the actual response at time t; p is the confidence level; $\tilde{F}_{\beta}(t, P)$ is the estimated value of CVaR, namely the energy storage capacity required to coordinate with a VPP. In the case of discrete distribution, m is the number of simulated data; a is the value of VaR; β is the confidence level.

The other two constraints need to be mentioned. On the one hand, there is an upper limit on the energy storage capacity to supplement the VPPs. On the other hand, when the risk requirements are met, the energy storage capacity should be over-prepared and scheduled to find the best energy storage capacity to improve efficiency and economy. Equations (11) and (12) can be further obtained from Eq. (9).

$$E_{\lambda_1, battery} = a + \frac{1}{m \cdot \lambda_1} \sum_{k=1}^{m} \left[f\left(t, P^k\right) - a\right]^+ \tag{11}$$

$$E_{\lambda_2,battery} = a + \frac{1}{m \cdot \lambda_2} \sum_{k=1}^{m} \left[f\left(t, P^k\right) - a \right]^+ \tag{12}$$

where $E_{\lambda_1,battery}$ is the efficient energy storage capacity; $E_{\lambda_2,battery}$ is the promised energy storage capacity. λ_1 and λ_2 reflect the error range of short-term response of a VPP. λ_1 is the critical value of not being punished by the market and being punished by the market, and λ_2 is the critical value of capacity of a VPP to participate in the market.

4 Benefits of VPPs Participating in Auxiliary Service Market Coordinating with Energy Storage Allocation

In general, the revenue from the auxiliary service market comes partly from the service revenue of the VPPs subject, and partly from the supplementary service revenue of the energy storage system. It also means that if there is no energy storage coordinating the VPPs' operations, then there will be a partial loss of profit, namely the opportunity loss mentioned in this chapter.

4.1 Benefits of VPP Participating in Auxiliary Services

VPPs are compensated for services to complete transactions and provide auxiliary services. This section is modeled with the north China auxiliary service market operating rules for an interval of 15 minutes, with the revenue formula per interval and cumulative full-day revenue as shown in Eqs. (13) and (14).

$$F_t = K^t \times \min\left\{ \frac{E_t}{P_{base}^t}, 1 \right\} \times \min\left\{ E_t, P_{base}^t \right\} \times t \times A_{out}^t \tag{13}$$

$$W = \sum_{t \in T} F_t \tag{14}$$

where F_t is the valley filling auxiliary revenue obtained by market entities at time t; K^t is the market coefficient indicating the reciprocal of the average load rate of thermal power units in the provincial power network, and is temporarily set as 1.2 in this part. $\min\left\{ \frac{E_t}{P_{base}^t}, 1 \right\}$ is the contribution rate of valley filling; E_t is the peak-adjusting power of response at time t; P_{base}^t is the benchmark valley filling power at time t; t is the response time, which is with a time interval of 15 min in this part. A_{out}^t is the market clearing price at time t; W is the total revenue of daily ancillary services; T is the opening session for ancillary services.

The incremental investment and maintenance costs of VPPs providing auxiliary services are a small part compared to the initial construction investment and costs, so the incremental part is not discussed in this chapter. However, the opportunity loss factor is included. If the actual response situation and the expected response situation deviate and exceed the provisions of the market operation rules, such losses will occur as shown in Eq. (15).

$$A_{e,t} = \gamma\left(K^t \times E_t \times t \times A_{out}^t\right)$$
$$\begin{cases} \gamma = 0, a < \lambda_1 \\ \gamma = 1 - \dfrac{E_t}{P_{base}^t}, \lambda_1 \le a < \lambda_2 \\ \gamma = 1, a \ge \lambda_2 \end{cases} \tag{15}$$

where $A_{e,t}$ is the penalty cost at time t and γ is the penalty coefficient. $a = \frac{E_t - P_{base}^t}{E_t}$ is the load deviation rate.

4.2 Benefits of Energy Storage

In order to avoid the deviation situation that the actual valley filling load of VPPs can not reach the expected load, the energy storage system needs to provide additional capacity to replenish the load. At the same time, energy storage can use this load to provide the required power to the grid during peak hours, so as to achieve the spread gains beyond the market income. The different types of energy storage resource revenue structure are shown in Fig. 2.

In Fig. 2, $E_{b,battery}$ is the total capacity of the energy storage equipment owned by the aggregator. $w_{u,t}$ refers to the income of the energy storage directly participating in the auxiliary service at time t. $W_{u,f}$ is the income of energy storage resources when they discharge in non-market period under the peak-valley electricity prices. W_u is the total revenue of energy storage participating in the auxiliary service market. $w_{l,t}$ is

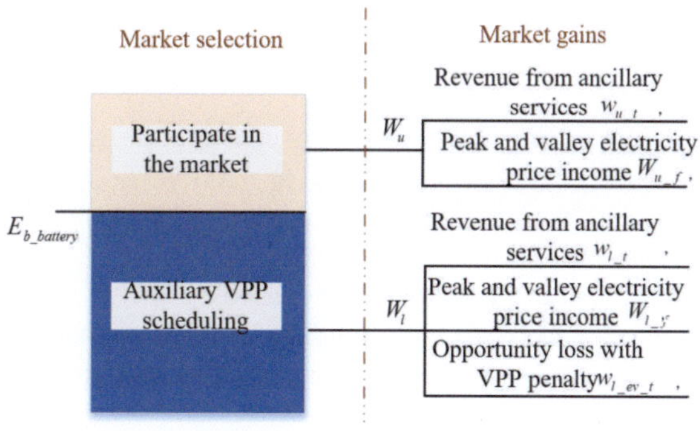

Fig. 2 Strategies of energy storage coordination VPPs participating in auxiliary services

the market income of coordinated energy storage participation in auxiliary services. $W_{l,f}$ refers to the peak and valley electricity price income for coordinated energy storage. $w_{l,ev,t}$ is the opportunity loss reduced after coordinating the energy storage for assisting VPPs. W_l is the total benefit of coordinated energy storage.

For energy storage resources directly involved in auxiliary services, the formulation for calculating the direct revenue of auxiliary services is shown in Eq. (16). The revenue of peak-valley tariff difference is shown in Eq. (17) and the total revenue of auxiliary services market is shown in Eq. (18).

$$w_{u,t} = K^t \times \min\left\{\frac{E_{t,b}}{P^t_{base,b}}, 1\right\} \times \min\left\{E_{t,b}, P^t_{base,b}\right\} \times t \times A^t_{out} \tag{16}$$

$$W_{u,f} = \left(p_p - p_v\right) \times Q_u \tag{17}$$

$$W_u = W_{u,f} + \sum_{t \in T} w_{u,t} \tag{18}$$

where $E_{t,b}$ is the actual response capacity of the energy storage resource at time t. $P^t_{base,b}$ is the reported response capacity of the energy storage resource at time t. p_p and p_v are energy storage discharge price and charging price. $Q = Q_u + Q_l$ is the total energy storage capacity participating in the auxiliary service market.

For the energy storage resources that coordinate VPPs participating in auxiliary service market, the revenue of their auxiliary service market is shown in Eq. (19). The revenue of peak-valley tariff difference is shown in Eq. (20). The opportunity loss reduction of coordinated energy storage is calculated in Eq. (21) and the total revenue of coordinated energy storage is shown in Eq. (22).

$$\begin{aligned} w_{l,t} &= K^t \times \min\left\{\frac{\Delta E_t + E_t}{P^t_{base}}, 1\right\} \times \min\left\{(\Delta E_t + E_t), P^t_{base}\right\} \times t \times A^t_{out} - F_t \\ &= K^t \times \frac{2E_t \cdot \Delta E_t}{P^t_{base}} \times \Delta E_t \times t \times A^t_{out} \end{aligned} \tag{19}$$

$$W_{l,f} = \left(p_p - p_v\right) \times Q_l \tag{20}$$

$$w_{l,ev,t} = \gamma\left(K^t \times E_{t,b} \times t \times A^t_{out}\right) \tag{21}$$

$$W_l = W_{u,f} + \sum_{t \in T}\left(w_{u,t} + w_{l,ev,t}\right) \tag{22}$$

where ΔE_t is the load growth caused by energy storage regulation. Q_l is the energy storage capacity for coordinating VPPs participating in auxiliary services and $E_{t,b}$ is the auxiliary adjustment capacity of energy storage at time t.

In the calculation of energy storage configuration cost, this part uses the annual value calculation equation of the battery storage system as the energy storage cost measurement method, then the annual cost of energy storage is calculated as shown in Eq. (23).

$$A_b = (P_{bess} \times \alpha + E_{bess} \times \beta) \times \frac{(1+i)^{T_{life}} \times i}{(1+i)^{T_{life}} - 1} \tag{23}$$

where A_b is the annual cost of the energy storage system and P_{bess} is the planned value of energy storage power. α is the amortization of power cost for energy storage and E_{bess} is the planned value of energy storage capacity. β is the amortization cost of storage capacity. i is the level of interest rates and T_{life} is the service life of energy storage. The calculation method of some important parameters is as follows.

$$\beta = \frac{C_E}{T_{life}} + C_m \tag{24}$$

$$\alpha = \lambda \cdot \beta \tag{25}$$

$$T_{life} = 1 \bigg/ \sum_{i=1}^{N} \frac{1}{K_{eye,D(i)}} \tag{26}$$

$$L_{eye,D} = 694 D^{-0.795} \tag{27}$$

where C_E is the investment cost of the energy storage system and C_m is the maintenance cost of the energy storage system. λ is the ratio of power investment to capacity investment of the energy storage system. $K_{eye,D(i)}$ is the battery life when the charge and discharge depth is $D_{(i)}$. $L_{eye,D}$ is the battery life when the charging and discharging depth of the lithium battery is D.

4.3 *Model for Optimal Allocation of Energy Storage Based on Maximizing Market Benefits*

The objective function (28) is established to maximize the annual market benefits of a VPP in the auxiliary service market.

$$\max \text{Pro} = \sum_{d=1}^{365} \left(W^d - \sum_{t \in T} A_{e,t}^d + W_u^d + W_l^d \right) - A_b \tag{28}$$

where d is the date count; pro is the calculation equation of the aggregator's annual profit. And optimize the energy storage capacity through optimization operation (29).

$$Q = Q_u + Q_l \tag{29}$$

Considering that the ability of energy storage to charge and discharge energy will be limited by the rated power of energy storage charging and discharging, the constraint of energy storage power is set as shown in Eq. (30).

$$\max_{d\in(1,365),t\in T}\left\{E_t^d\right\}\cdot(\lambda_1-\lambda_2)\leq P\cdot\Delta t \tag{30}$$

5 Case Studies

One of the classic scenarios in VPPs [16], a VPP aggregated with EVs is introduced in this section. The capacity of the VPP participating in the auxiliary service market and the revenue under a certain confidence are calculated with the market prices and market coefficient of the auxiliary service market. Note that the impact of the long-term planning of the energy storage system on the response capacity is excluded.

5.1 Scenario Introduction

To determine the coordinated capacity of energy storage requires first determining the responsiveness of the VPP aggregated by EVs, and then simulating the level of deviation in providing auxiliary services for participating in market transactions. In this case, the distribution function of short-term market participation will be fitted, and then the annual response capacity will be simulated by Monte Carlo simulation method. Finally, the LSSVM method is used to obtain the distribution function of the response deviation through the above statistical samples.

In the fitting process, the response size of the auxiliary service market is randomly sampled for 100 days to fit the distribution function. The fitting results show that the short-term response capacity size obeys a normal distribution with a mean of 368.88 and a variance of 24.34. Based on this normal distribution function, combined with the normrnd function in MATLAB, the capacity of the VPP participating in the market for 365 days was simulated, and the results are shown in Fig. 3.

The LSSVM simulation was used to calculate 960 intra-day load curves, and the distribution of response deviations is shown in Table 3.

Then, according to the load simulation results, the CVaR method with 90% confidence is used to measure the daily peaking reserve storage capacity, and the distribution of the measurement results is shown in Figs. 4 and 5. Among them, Fig. 4 shows the result of benefit energy storage allocation and Fig. 5 shows the result of promised energy storage allocation. To reflect the distribution relationship, Figs. 4 and 5 are depicted by normal distribution probability diagram.

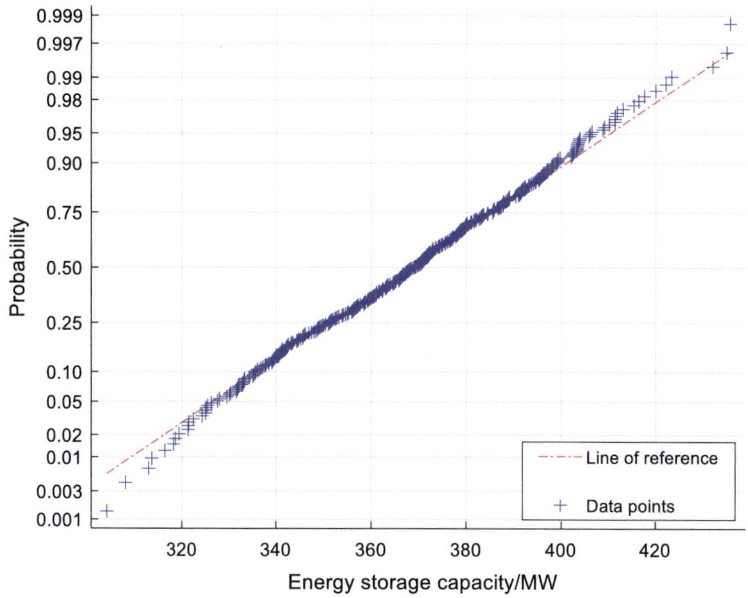

Fig. 3 Fitting results of a VPP aggregated with EVs

Table 3 The deviation distribution between reported capacity and actual response capacity

Deviation interval	Count results	Probability distribution
0~0.01	245.00	0.26
0.01~0.02	260.00	0.27
0.02~0.03	168.00	0.18
0.03~0.04	109.00	0.11
0.04~0.05	68.00	0.07
0.05~0.06	50.00	0.05
0.06~0.07	27.00	0.03
>0.07	33.00	0.03

Setting the auxiliary service market error range is not more than 5%, the auxiliary service clearing price is 300 yuan/MWh. Take the peak and valley electricity price difference is 1030 yuan/MWh, and the peak and flat section electricity price difference is 500 yuan/MWh. The investment cost is 5,166,000 yuan/MWh and the maintenance cost is 1,74,000 yuan/(MWh·year). The cost ratio λ is 1.172, and the market interest rate level is 8%. Accordingly, the following simulation results are obtained.

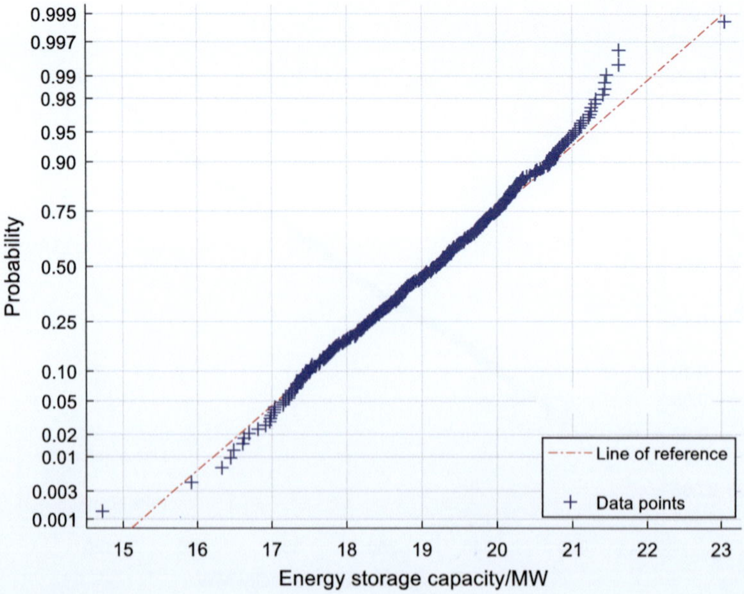

Fig. 4 Scale of benefit energy storage

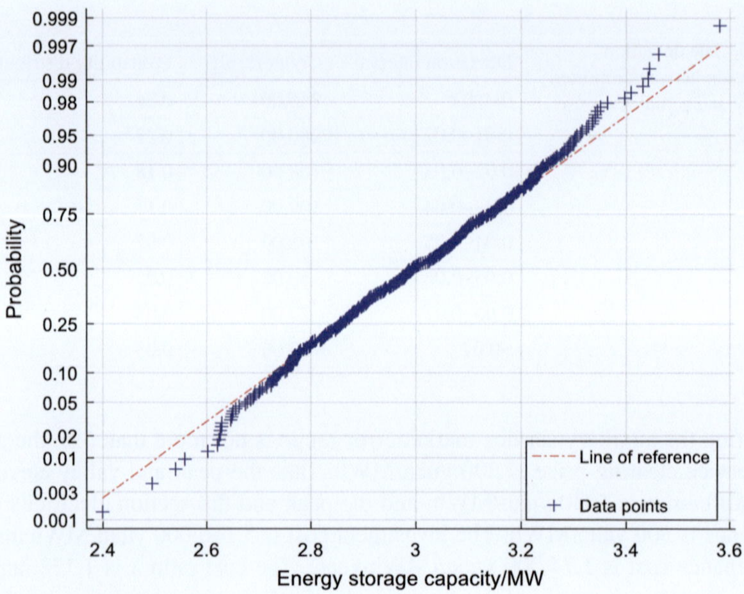

Fig. 5 Scale of promised energy storage

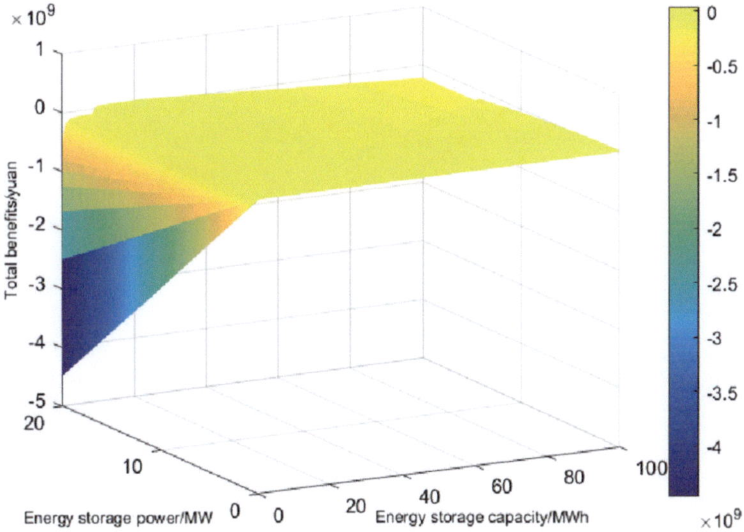

Fig. 6 The benefits of different energy storage power and capacity configuration results

5.2 *Simulation Results*

(1) Optimal energy storage configuration

According to the simulation results of the above scenario, the benefits of different energy storage power and capacity configuration results where the confidence level is taken as 90% are shown in Fig. 6. Figure 7 shows the partial display of the results.

In the scenario set up in this section, where no energy storage is configured, the annual benefit of the VPP is 33.97 million yuan. At the confidence level of 90%, the energy storage capacity configuration is 13.66 MWh. Moreover, the energy storage power configuration is 1.96 MW, at which time the annual benefit of the VPP and energy storage joint dispatch is 39.15 million yuan, and the incremental benefit reaches 5.18 million yuan.

In the scenario optimization, power investment in energy storage systems does increase the response accuracy, but the total investment cost will inevitably rise, which will cause the problem of excessive investment and the waste of energy storage resources. The negative impact will offset the effect of power investment in improving the efficiency of VPPs scheduling, from another perspective. Similarly, the rapid increase in the cost of excessive energy storage capacity allocation prevents the revenue from participating in the market from covering the cost, and thus generates no significant incremental benefits, which is fully demonstrated in Fig. 7. Therefore, the energy storage system must play its role in coordinating with the operation, and deeply play its scheduling value.

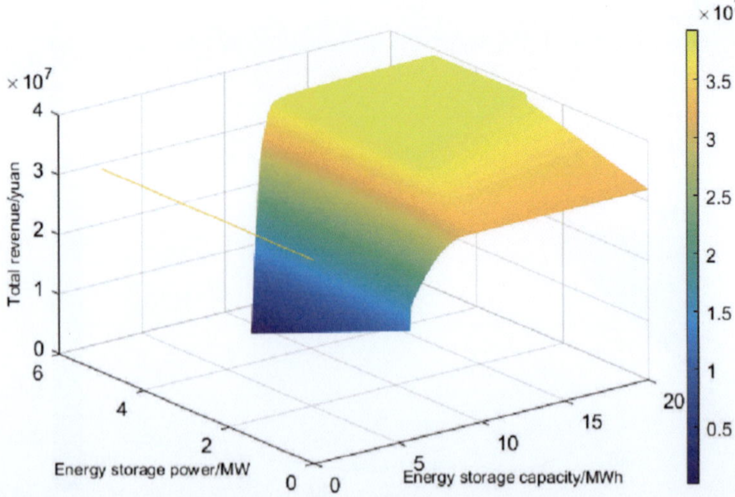

Fig. 7 A local view of the result

(2) Energy storage construction schemes under different confidence levels

Under different CVaR confidence levels, the corresponding optimal energy storage construction plan is calculated, and its final results are shown in Table 4.

As can be seen from Table 4, as the confidence level requirements increase, the requirements for energy storage capacity and power scale also increase. This is due to the fact that the higher the reliability requirements for VPPs aggregated with EVs to energy storage coordination and dispatch, the more it costs and therefore more energy storage resources are needed.

(3) Typical daily market conditions under the optimal energy storage scale

In the case of optimal energy storage scale, based on the relevant parameters in Sect. 5.1, the operation of the VPP and energy storage on a typical day under this energy storage configuration is simulated. The reported capacity, load of the VPP, load share of energy storage, coordinated dispatch capacity and opportunity loss are measured. The measured results of the auxiliary service market operation on a typical day are shown in Table 5.

After typical daily market operations, the energy storage system timely conducts the charging and discharging process according to the instructions, and the effect of reducing the response deviation of VPPs is obvious, thus significantly increasing the overall service effect and benefits of VPPs. Overall, the coordinated response volume of the energy storage system plays two roles, namely, improving the response scale and reducing the opportunity loss caused by the stochastic volatility of the DERs.

The analysis of the regulation effect of the VPP participating in the auxiliary service market is carried out with the energy storage resources on a certain day. According to the optimized allocation results, 12.40 MWh of energy storage capacity

Table 4 Optimal allocation scheme and storage efficiency under different confidence levels

Confidence level	Total benefits (thousand yuan)	Energy storage capacity (MWh)	Energy storage power (MW)
0.05	41.70	1.03	0.12
0.1	41.63	1.88	0.18
0.15	41.55	2.22	0.23
0.2	41.45	2.64	0.31
0.25	41.38	3.02	0.4
0.3	41.70	1.03	0.12
0.35	40.90	4.00	0.58
0.4	41.05	4.7	0.68
0.45	40.54	5.19	0.75
0.5	40.71	5.89	0.85
0.55	40.54	6.45	0.93
0.6	40.67	7.22	1.04
0.65	40.29	7.85	1.13
0.7	40.40	8.62	1.24
0.75	40.37	9.46	1.36
0.8	39.66	10.37	1.49
0.85	39.77	11.77	1.69
0.9	39.15	13.66	1.96
0.95	38.47	18.56	2.66

Table 5 Simulation results of typical daily market operations

The total capacity (MW)	Coordinate energy storage capacity (MW)	Ratio of VPP load (%)	Ratio of energy storage load (%)	Ratio of capacity that the VPP adjusted (%)	Compensated opportunity loss (yuan)
372.25	12.40	96.7	3.3	18.5	4518.44
364.07	9.12	93.4	6.6	20.6	4902.54
379.81	13.65	94.7	5.3	19.4	4826.12
354.54	8.81	95.2	4.8	22.8	5305.33
371.80	12.38	93.4	6.6	17.0	4139.21
350.05	8.66	94.4	5.6	20.9	4793.38

out of 13.66 MWh is applied to this coordinated and optimized dispatch, and 1.26 MWh of energy storage capacity is directly involved in the auxiliary service market. The final market operation results are shown in Fig. 8, which shows the VPP regulation and the regulation effect of energy storage during the auxiliary service market open.

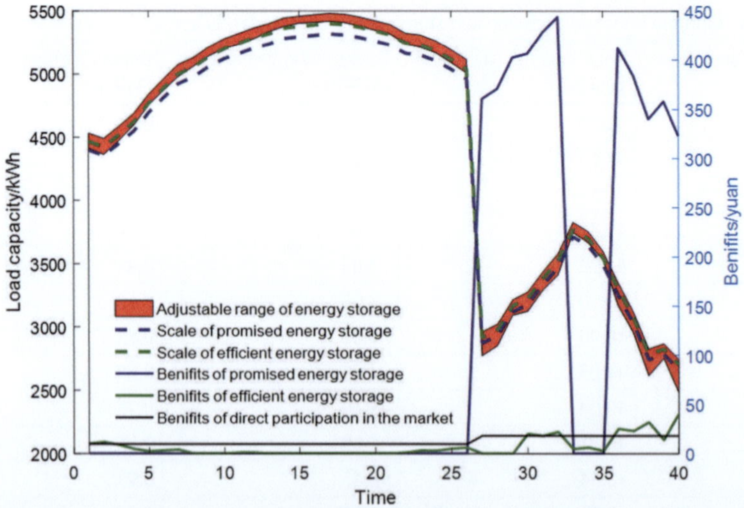

Fig. 8 Operation results of the VPP participating in auxiliary service market on a typical day

Figure 8 vividly shows the time change characteristics of load capacity when the VPP makes market transactions. Early morning is sufficient responsive period of the VPP, accordingly, only a small amount of benefit energy storage supplementation is required. However, during the noon and afternoon periods (12:30–18:00), the uncertainty load impact resulting from the stochasticity of DERs is very large, requiring not only benefit energy storage but also sufficient supporting energy storage.

In this process, the total benefit is 51,259.73 yuan, among which, the incremental benefit generated through the coordinated scheduling of energy storage resources is 5008.15 yuan, including 272.80 yuan of opportunity loss made up by benefit energy storage and 4229.49 yuan of opportunity loss made up by promised energy storage. Moreover, the benefit made up by direct energy storage capacity is 505.86 yuan. Promised energy storage plays a more obvious role in the coordination and scheduling of auxiliary services.

6 Conclusion

According to the analysis of the previous sections, this chapter studies the trading mechanism of VPPs' collaborative energy storage system operation jointly participating in the auxiliary service market, which focuses on how the energy storage system allocates capacity to achieve the optimal service effect with VPPs, so as to maximize the market revenue. It also provides the case analysis of the transaction and operation situation of VPPs participating in the auxiliary service market, and verifies the series of theories and methods proposed in this chapter. The main conclusions are as follows:

(1) At present, the auxiliary service market is gradually opening to the third-party market entities, and the access conditions are becoming more and more relaxed, so that the market entities formed by DERs aggregation can obtain a considerable part of the income through market transactions. Considering the best operation solutions for energy storage to coordinate with VPPs, including scheduling plans and bidding strategies, it can make such VPPs obtain rich return.
(2) The bidding strategy for VPPs coordinated with energy storage in the auxiliary service market is proposed, and the CVaR method is used to calculate the storage resource capacity required for short-term response capacity of VPPs, which can ensure the maximum return.
(3) The supplementary capacity of the energy storage system and the revenue with the VPPs are measured from the long-term and short-term dimensions, respectively. Short-term response capacity is predicted by big data technology, and long-term response capacity is obtained by statistical analysis of samples with short-term actual response capacity.
(4) The random volatility carried by DERs themselves causes a response deviation of VPPs in providing auxiliary services. This chapter proposes that the bidding strategy of VPPs collaborative energy storage system operation can effectively make up for this response deviation and improve the operation effect. And it can eliminate the opportunity loss caused by deviation, cover the operating costs, thus getting higher profit.

Furthermore, the extent of the loss of benefit resulting from the VPPs participating in auxiliary service market transactions depends on the deviation between the reported capacity and the actual responsiveness. The fundamental way to solve this problem is to improve the capacity prediction accuracy of VPPs and improve the regulation technology, which will greatly improve the operational effectiveness of VPPs. The market performance risk cost of VPPs at different confidence levels will be the direction of further research as well.

References

1. Y. Chen, J. Zhao, J. Ma, Fast decoupled multi-energy flow calculation for integrated energy system. J. Mod. Power Syst. Clean Energy **8**(5), 951–960 (2020)
2. Q. Zhao, Y. Shen, M. Li, Control and bidding strategy for virtual power plants with renewable generation and inelastic demand in electricity markets. IEEE Trans. Sustain. Energy **7**(2), 562–575 (2016)
3. Y. Chen, Y. Yao, Y. Lin, X. Yang, Dynamic state estimation for integrated electricity-gas systems based on Kalman filter. CSEE J. Power Energy Syst. **10**, 1–11 (2020)
4. Z. Yi, Y. Xu, H. Wang, L. Sang, Coordinated operation strategy for a virtual power plant with multiple DER aggregators. IEEE Trans. Sustain. Energy **12**(4), 2445–2458 (2021)

5. J. Naughton, H. Wang, M. Cantoni, P. Mancarella, Co-optimizing virtual power plant services under uncertainty: a robust scheduling and receding horizon dispatch approach. IEEE Trans. Power Syst. **36**(5), 3960–3972 (2021)
6. Z. Yi, Y. Xu, J. Zhou, W. Wu, H. Sun, Bi-level programming for optimal operation of an active distribution network with multiple virtual power plants. IEEE Trans. Sustain. Energy **11**(4), 2855–2869 (2020)
7. J. Valinejad, M. Marzband, M. Korkali, Y. Xu, A.S. Al-Sumaiti, Coalition formation of microgrids with distributed energy resources and energy storage in energy market. J. Mod. Power Syst. Clean Energy **8**(5), 906–918 (2020)
8. J.-F. Toubeau, J. Bottieau, Z. De Grève, F. Vallée, K. Bruninx, Data-driven scheduling of energy storage in day-ahead energy and reserve markets with probabilistic guarantees on real-time delivery. IEEE Trans. Power Syst. **36**(4), 2815–2828 (2021)
9. B. Xu et al., Scalable planning for energy storage in energy and reserve markets. IEEE Trans. Power Syst. **32**(6), 4515–4527 (2017)
10. R. Zhang, B. Hredzak, J. Fletcher, Dynamic aggregation of energy storage systems into virtual power plants using distributed real-time clustering algorithm. IEEE Trans. Industr. Electron. **68**(11), 11002–11013 (2021)
11. W. Tang, H. Yang, Optimal operation and bidding strategy of a virtual power plant integrated with energy storage systems and elasticity demand response. IEEE Access **7**, 79798–79809 (2019)
12. M. Giuntoli, D. Poli, Optimized thermal and electrical scheduling of a large scale virtual power plant in the presence of energy storages. IEEE Trans. Smart Grid **4**(2), 942–955 (2013)
13. T. Kovaltchouk, A. Blavette, J. Aubry, H.B. Ahmed, B. Multon, Comparison between centralized and decentralized storage energy management for direct wave energy converter farm. IEEE Trans. Energy Convers. **31**(3), 1051–1058 (2016)
14. B. Fang, X. Ren, W. Lin, H. Li, G. He, W. Zhou, Analysis of the influence of large-scale integration of centralized energy storage into the power grid on voltage security and stability of power system. In *2021 3rd Asia Energy and Electrical Engineering Symposium (AEEES)* (2021), pp. 759–764
15. Q. Ge et al., Industrial power load forecasting method based on reinforcement learning and PSO-LSSVM. IEEE Trans. Cybern. **52**(2), 1112–1124 (2022)
16. D. Liu, L. Wang, M. Liu, H. Jia, H. Li, W. Wang, Optimal energy storage allocation strategy by coordinating electric vehicles participating in auxiliary service market. IEEE Access **9**, 95597–95607 (2021)

Chapter 7
Dynamic Pricing Strategy of Virtual Power Plants Based on DDPG Reinforcement Learning Algorithm

The optimal operation of VPPs should be achieved by aggregating more DERs and improving the responsiveness of DERs. The design of a reasonable incentive mechanism and a pricing model can improve the operating revenues of VPPs, and also increase the benefits of DERs from providing services for power systems, thus promoting the awakening responsiveness.

Since fixed pricing strategies for VPPs are probably not effective enough in dealing with complex trading patterns, it is necessary to design a dynamic pricing model. A dynamic pricing strategy not only reflects the supply and demand contrasts to the trading parties in a timely manner in the electricity market, but also stimulates the economic value of VPPs. This chapter considers the economic potential of VPPs, with the objective of maximizing the combined benefits in the framework of peak-valley tariffs, demand-side response and spot markets. It is aimed to design VPPs that aggregate a large number of electric vehicles (EVs) as a classic scene. The focus is on the development of a dynamic time-sharing pricing model, which is based on deep deterministic policy gradient (DDPG) reinforcement learning algorithm. Finally, a case study is presented that the differences in revenue between peak-valley time-of-use tariff and hourly pricing strategies are compared. It is demonstrated that by increasing the frequency of pricing signals and quickly updating market prices, such strategies are more effective in guiding electricity consumption behaviour, exploiting the economic potential of VPPs and maintaining the stability of the grid.

1 Introduction

The wholesale electricity market is established with high entry conditions. Therefore, various DERs have to be aggregated to get access. For example, EVs are aggregated into VPPs [1], where the charging and discharging behaviours of aggregated EVs are guided according to market transactions and electricity prices, etc., to increase the revenue of VPPs. Scheduling alone does not directly and efficiently control the

H. Jia et al., *Business Models and Reliable Operation of Virtual Power Plants*,
https://doi.org/10.1007/978-981-19-7846-3_7

aggregated resources, and price-adjusted coordination approaches [2, 3] have been noted. VPPs need to process information such as market demand and set prices at different stages, and then the DERs determine the optimal response based on a value function with the efficient operation of the electricity market and the VPPs.

VPPs are not extremely active in the electricity market and the offers are relatively fixed, making it difficult for VPPs to adapt to the requirements of market. In order to fully stimulate the interest of VPPs to participate in trading, fixed offers can be replaced by periodic dynamic pricing to increase the frequency of offers, thus improving the transaction frequency. The dynamic design is a classical method for building real-time pricing models [4, 5] meeting the high-dimensional and high frequency pricing needs, where a large amount of computing resources is also necessary. Moreover, dynamic planning cannot eliminate the effects of instability caused by DERs.

Therefore, there is necessity to design pricing models applicable to VPPs, taking into account the response effect of DERs and comprehensive benefits. There are currently four pricing schemes: (1) cost analysis [6–8], where a percentage of profit is added to the cost of electricity purchase to form the declared electricity price; (2) DER behaviour analysis [9–11], where VPPs are priced based on the predicted elasticity coefficient of the price for DERs; (3) game theory [12, 13], that is, building a game model and finding the model equilibrium to determine VPPs bidding strategy; (4) intelligent optimization algorithms, such as competitive co-evolution algorithm [14, 15], fuzzy self-adaptive search algorithm [16] and reinforcement learning (RL) method [17, 18], etc.

For the above pricing strategies, cost analysis is the basic method, which is simple and widely applicable, but the crude analysis is difficult to maximize comprehensive returns and does not take into account external price information. Behaviours analysis of DERs to obtain elasticity coefficients or probability densities is also a classical method, but also does not accurately fit the daily dynamic load of DERs. The game theory method is effective as a solution to the symmetrical information problem in a fixed environment, but has shortcomings for the asymmetric information problem in a variable environment. All these traditional optimization methods have computing difficulties in dynamic pricing.

The RL method is a data-driven control method with an "action-to-reward" mechanism that allows for flexible changes in the system environment to real-time optimal control in response. Its model-free characteristic liberating the system modelling process [19] and reducing the complexity of modelling. Traditional RL method based on value function selects the action with the highest return as the strategy by comparing the size of different environment-action pairs. Therefore, the discrete processing of control actions, the method and degree of discrete processing are important, while the complexity of action space will increase geometrically with the increase of actions, limiting its use in complex decision problems. Deep deterministic policy gradient (DDPG) algorithm based on actor-critic system is another kind of RL algorithm, where the agent learns the mapping function between environment and action without discretization of the control action [18]. DDPG has a wider range of application scenarios to analyse dynamic pricing problems.

Therefore, this chapter provides methods of the price declaration of VPPs with DERs' participation in multiple electricity markets, tracking 15-min price information, with the historical data basis of DERs' behaviours, by DDPG algorithm to build a quarter-hourly dynamic pricing model in the real-time continuous strategy space. Comparing market returns at different pricing frequencies can verify the pricing effectiveness and function of the DDPG algorithm.

2 Analysis of Market Trading of VPPs

The electricity market has become increasingly tradable and active currently, but it has also resulted in inter-temporal power differentials and volatility. Demand response (DR) is becoming an important market participant, acting as a significant force in maintaining the reliability of power systems.

By widening the peak-valley tariffs, the average peak-valley time-of-use tariffs can reach about 3 times. Conducting DR transactions at specific times can add value to VPPs. The emergence and development of spot market has eliminated the traditional deviation, and the dynamic pricing of spot markets can more accurately compensate for forecast deviations.

A typical scenario of a VPP interacting with different DERs is shown in Fig. 1. This framework consists of three parties, the power system operator, the VPPs and the DERs, where the DERs can be represented by aggregated EVs. Meanwhile, the system operator optimizes the consumption behaviour of DERs to reduce the peak and valley gap. By peak shaving and valley filling, VPPs compensate for the power deviation caused by DR in the spot market.

Moreover, the DDPG algorithm operates on the basis of real-time external market price information and the behaviour information of DERs, and the algorithm will output a dynamic pricing strategy that enables VPPs to trade with power systems and DERs. Revenue from VPPs consists of three components: (1) DR revenue from controlling the orderly electricity consumption behaviour of DERs and reducing peak-valley differentials. (2) revenue from eliminating response bias in the spot market (3) service fee while purchasing time-of-use electricity for DERs.

Next, an iterative DDPG algorithm model is developed for dynamic pricing, with the participating market types being DR market and spot market. The revenue consists of market revenue and tariff revenue.

(1) DR market: The DR market conducts valley filling trade from 0:45 to 7:00 and peak trading between 12:45 and 16:00 every day. VPPs take the difference between the reference load curve and the actual load curve confirmed by its own predicted electric characteristics. The benefit of a VPP in DR market in formulated in (1).

$$R^a = \sum_{i=3}^{28} \left(Q_i^{Acutal} - Q_i^{Baseline}\right) p_i^a t + \sum_{i=51}^{64} \left(Q_i^{Baseline} - Q_i^{Acutal}\right) p_i^a t \quad (1)$$

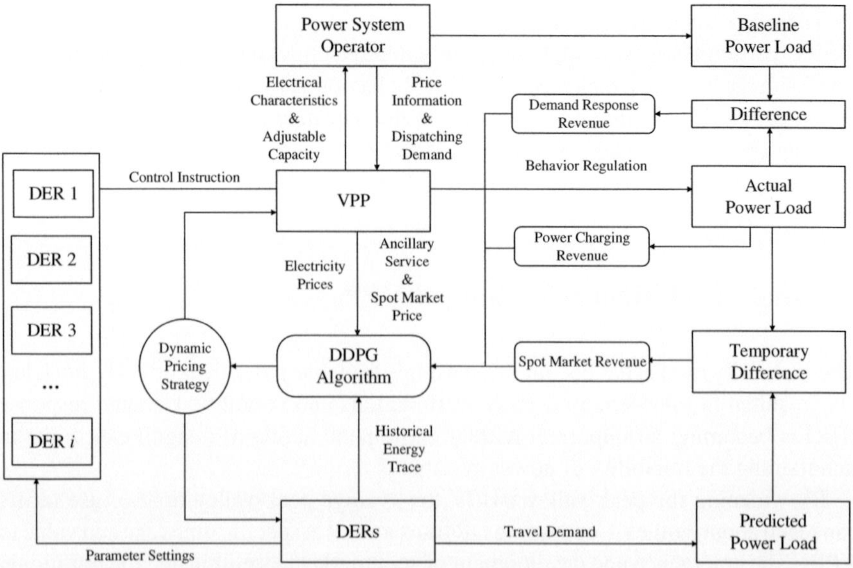

Fig. 1 Interaction framework of the proposed market trading of VPPs

where R^a is the benefit of a VPP in DR market. Q_i^{Acutal} is the real load value and $Q_i^{Baseline}$ is the benchmark load value at time point i. p_i^a is the DR clearing price at time point i and t is the time interval.

(2) Spot market: The deviation value between the predicted load curve and the actual curve at each time point of a VPP needs to be eliminated in the spot market according to DERs' power consumption and contract information. The benefit of a VPP from the spot market is calculated in (2).

$$R^s = \sum_{i=1}^{96} \left(Q_i^{Predict} - Q_i^{Actual}\right) p_i^s t \tag{2}$$

where R^s is the benefit of a VPP from the spot market. $Q_i^{Predict}$ is the expected load value time point i and p_i^s is the clearing price in the spot market at time point i.

(3) Electricity fee income: The benefit of a VPP for electricity fee can be divided into two parts, including time-of-use electricity purchase fee and service fee. The time period of time-of-use electricity price is divided into the peak time period with 12:00–21:00, the normal period with 7: 00–12: 00 and 21: 00–24:00, as well as the valley period with 0: 00–7: 00. The calculation formula is in (3).

$$R^c = \sum_{i=1}^{28} Q_i^{Actual}\left(p_i^c + p^L\right)t + \left(\sum_{i=28}^{48} Q_i^{Actual} + \sum_{i=84}^{96} Q_i^{Actual}\right)$$

$$\left(p_i^c + p^M\right)t + \sum_{i=48}^{84} Q_i^{Actual}\left(p_i^c + p^H\right)t \quad (3)$$

where R_c is the electricity fee income. p_i^c is the dynamic service fee at time point i. p_i^H, p_i^M, and p_i^L are the electricity purchase price during the peak period, the normal period, and the valley period, respectively.

The electricity consumption behaviour of DERs are diverse, and incentives such as changing the bidding prices of VPPs and differential service fees can be used to make DERs adjust electricity consumption time and capacity spontaneously, thus completing the optimal operation of VPPs and increasing revenue.

3 Principle of DDPG Reinforcement Learning Algorithm

The "action-to-reward" mechanism of RL refers to the guiding mechanism where the subject and the environment influence the subject's behavior by rewarding and punishing information during the interaction process. DDPG is an actor-critic based algorithm in RL [19]. The central view of this algorithm enables the use of actor network to form agent behavior and strategies. The Critic network can judge the quality of actions and guide the updating direction of actions [19].

As shown in Fig. 2, the DDPG structure contains an actor network with parameter θ and an critic network with parameter ω, which are brave to calculate deterministic strategy $\pi(s|\theta)$ and action value function $Q(s, a|\omega)$, respectively. Since the single network learning process is unstable, it draws on the successful experience of Deep Q-Network (DQN) fixed target network. Moreover, actor network and critic network are subdivided into one reality network with one target network respectively. The real network and the target network structure is the same, while parameters of the target network is utilized to determine the frequency by the soft update of real network parameters.

The visual environment of each step of the DDPG algorithm is the market price information obtained from VPPs at each time point, and the load change information is obtained from DERs. After feeding the environment information into the actor network, the model outputs the dynamically priced action value quarter-hourly.

Through the continuous changing market prices, the income difference between the two pricing methods, the output dynamic pricing and fixed pricing, is used as the reward value. The external environment information is then introduced into the critical network to evaluate the effectiveness of the pricing strategy. As revenue increases, the pricing strategy works better. It should be noticed that network weights of the actor network are adaptively coordinated with feedback from the critic network.

According to the output price change information and its own value function, the DERs output the load change value under the new price information. Together with the new round of market price information received by VPPs, the environmental input value of the new step changes as well.

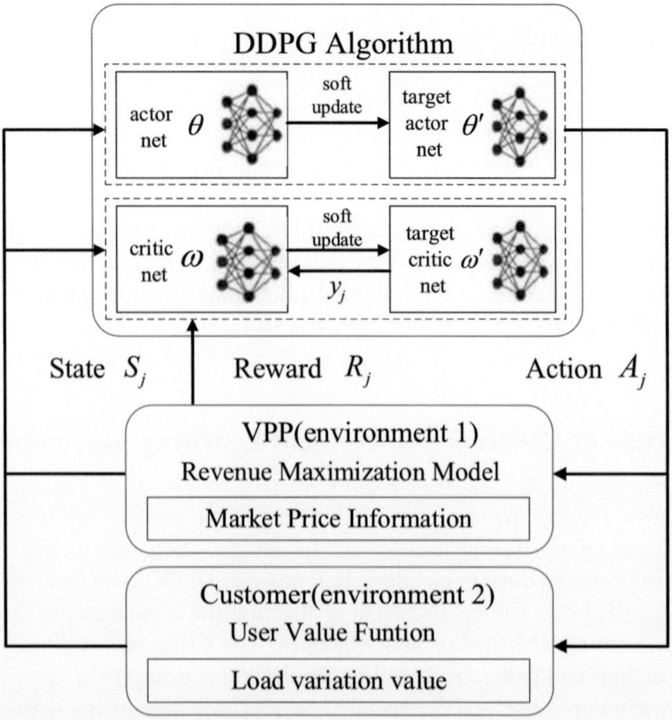

Fig. 2 Calculation step of DDPG algorithm

When the training reaches a certain number of rounds, the parameters of target net will be updated on a sliding average based on the actual value of the real network parameters.

The loss function of reality network is formulated in (4) [20].

$$J(\omega)=\frac{1}{m}\sum_{j=1}^{m}\left(y_j - Q(\phi(S_j), A_j, \omega)\right)^2 \tag{4}$$

In each iteration, m samples are sampled from the experience playback set D. The current target value is calculated as (5).

$$y_j = \begin{cases} R_j \; is\_end_j \; is \; true \\ R_j + \gamma Q\big(\phi(S_{j+1}), A_{j+1}, \omega\big) \; is\_end_j \; is \; false \end{cases} \tag{5}$$

where m is the number of samples, $Q\big(\phi(S_j), A_j, \omega\big)$ is the action A_j taken by the sample in the state, where the action value is calculated by the reality critic network. y_j is the target action value calculated by the target critic network. R_j is the reward obtained by the sample taking an action A_j in the state of S_j, and γ is the discount factor.

The loss gradient for a real actor network is formulated in (6).

$$\nabla J(\theta) = -\frac{1}{m}\sum_{j=1}^{m}(\nabla_a Q(s_i, a_i, \omega)) \tag{6}$$

where the gradient descent method is used to find the minimum value of the loss function $J(\theta)$. The loss function is equivalent to the process of maximizing the action value $Q(\phi(S_j), A_j, \omega)$.

At the end of each iteration, the critic target net and actor target net parameters are updated in the following manner.

$$\omega' \leftarrow \tau\omega + (1 - \tau)\omega' \tag{7}$$

$$\theta' \leftarrow \tau\theta + (1 - \tau)\theta' \tag{8}$$

If S_j is the termination state, the round is iterated out.

4 Dynamic Pricing Model for VPPs

4.1 *Price Elasticity Coefficient of DERs Based on Value Function*

Under the dynamic adjustment of multiple rounds of electricity market price, the electricity consumption behavior of DERs will be effectively guided by the price. This section examines the relationship between price changes and total load changes of DERs, and the relationship based on value function. By recording the load status of the DERs at different prices, to further characterize the parameters of the elasticity coefficient.

According to the Generalized Leontief value function, the total cost of a VPP can be expressed in (9) with the price and demand in different time periods.

$$C = g(q_i)\sum_{i=1}^{96}\sum_{j=1}^{96}\omega_{ij}\sqrt{(p_i p_j)} \tag{9}$$

where i, j represent different time periods. C represents the total cost of a VPP and $g(q_i)$ represents the demand of aggregated DERs. p_i and p_j respectively represent price in time period i and time period j.

The ratio σ_i of fixed period expenses to total expenses can be expressed in (10).

$$\sigma_i = \frac{\sum_j \omega_{ij}\sqrt{(p_i p_j)}}{\sum_{k=1}^{96}\sum_j \omega_{kj}\sqrt{(p_k p_j)}} \tag{10}$$

Thus, the self-elastic coefficient ε_{ii} and the mutual elastic coefficient ε_{ij} of the electricity consumption behavior of DERs can be calculated as (11) and (12).

$$\varepsilon_{ii} = \frac{1}{2}\left(\frac{\omega_{ii}}{\sum_{ki}\omega_{ik}}\sqrt{\frac{p_i}{p_k}} - 1\right) \tag{11}$$

$$\varepsilon_{ij} = \frac{1}{2}\left(\frac{\omega_{ij}}{\sum_{ki}\omega_{ik}}\sqrt{\frac{p_j}{p_k}}\right) \tag{12}$$

4.2 Dynamic Pricing Model Based on DDPG Reinforcement Learning

The significance and purpose of electricity price guidance are viewed from two aspects. One is the economic level, which is the income of VPPs, and the other is the dispatching level, which can suppress the load fluctuation of the power grid and reduce the peak and valley difference. Dynamic pricing strategies are established based on the revenue components mentioned above to reflect the profit results; the load changes of DERs and economic efficiency reflect the response effectiveness. The comprehensive benefit maximization function shows in (13).

$$\max R = \sum_i \left(R^a + R^s + R^c\right) i = 1, 2, \ldots, n \tag{13}$$

In order to ensure the overall rationality of dynamic pricing strategy and avoid the damage of excessive profit to DERs' benefit under greedy algorithm, the pricing model should meet the following constraints:

(1) The earnings of a single trading day under the dynamic pricing strategy should not be lower than the earnings under the fixed pricing strategy as formulated in (14)–(17).

$$R^{a'} = \sum_{i=3}^{28}\left(Q_i^O - Q_i^{Baseline}\right)p_i^a t + \sum_{i=51}^{64}\left(Q_i^O - Q_i^{Baseline}\right)p_i^a t \tag{14}$$

$$R^{s'} = \sum_{i=1}^{96}\left(Q_i^{Predict} - Q_i^O\right)p_i^s t \tag{15}$$

$$R^{c'} = \sum_{i=1}^{28} Q_i^O (p_0 + p^L)t + \left(\sum_{i=28}^{48} Q_i^O + \sum_{i=84}^{96} Q_i^O\right)(p_0 + p^M)t + \sum_{i=48}^{84} Q_i^O (p_0 + p^H)t \tag{16}$$

$$R_i^a + R_i^s + R_i^c \geq R_i^{a'} + R_i^{s'} + R_i^{c'} \; i = 1, 2, \ldots, n \tag{17}$$

where Q_i^O refers to the load in the period before implementing the dynamic pricing. $R^{a'}$, $R^{s'}$, and $R^{c'}$ respectively refers to the benefit from the DR market, the benefit from the spot market and the service fee income before implementing the dynamic pricing. p_0 refers to the fixed service fee before implementing the dynamic pricing.

(2) After the implementation of dynamic pricing, the total energy consumption of DERs should remain unchanged, as formulated in (18).

$$\sum_j \sum_i^{96} Q_i^O = \sum_j \sum_i^{96} Q_i^{Actual} \; j = 1, 2, \ldots, n \tag{18}$$

(3) After the implementation of dynamic pricing, the total service cost of DERs should not be higher than the service cost under the fixed service fee, as formulated in (19).

$$\sum_j \sum_i^{96} Q_i^O p_i^c t \leq \sum_j \sum_i^{96} Q_i^{Actual} p_0 t \; j = 1, 2, \ldots, n \tag{19}$$

(4) Considering the economic limitations from both DERs and power systems, the service fee should be varied within a certain range, as formulated in (20).

$$p_i \in (p_{\min}, p_{\max}) \tag{20}$$

4.3 Solution of Dynamic Pricing Model Based on DDPG Reinforcement Learning

The inputs of the DDPG algorithm are actor current net, actor target net, critic current net, and critic target net where corresponding parameters are θ, θ', ω, ω', attenuation factor γ, soft update coefficient τ, sample number of batch gradient descent m, target Q network parameter update frequency C, maximum number of iterations T, random noise function N.

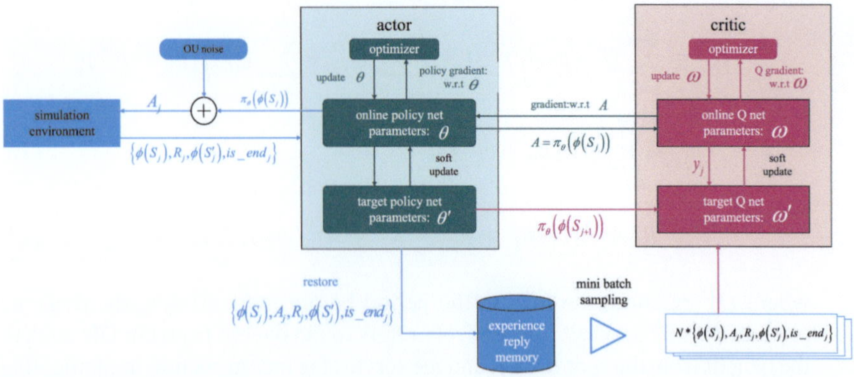

Fig. 3 The implementation framework for DDPG

The outputs of the DDPG algorithm are the best actor current net parameters θ and the critic current network parameters ω.

The implementation framework for DDPG is illustrated in Fig. 3. with the following steps.

Step 1: Random initialization $\theta, \omega, \omega' = \omega, \theta' = \theta$. Clear the set of experience replays D.

Step 2: Iterate from the first step of the first round.

(A) Initialize the initial state S of the current state sequence and get its feature vector $\phi(S)$.

(B) The current actor network $A = \pi_\theta(\phi(S)) + N$ gets an action based on the state S.

(C) Perform action A, and get new status S', reward R.

(D) Store the quintuple value $\{\phi(S), A, R, \phi(S'), is\_end\}$ into the experience playback set D.

(E) Set $S = S'$.

(F) Sample m samples from the experience playback set D and calculate the current target Q value y_j.

$$y_j = \begin{cases} R_j \, is\_end_j \, is \, true \\ R_j + \gamma Q(\phi(S_{j+1}), A_{j+1}, \omega) \, is\_end_j \, is \, false \end{cases} \tag{21}$$

(G) Use the mean square error loss function $\frac{1}{m}\sum_{j=1}^{m}\left(y_j - Q(\phi(S_j), A_j, \omega)\right)^2$ to update all parameters of the current network through gradient back propagation of the neural network.

(H) Use $J(\theta) = -\frac{1}{m} \times \sum_{j=1}^{m} Q(s_i, a_i, \theta)$ to update all parameters of the actor's current network through gradient back propagation of the neural network.

(I) If $T\%C = 1$, then update the target network and actor target network parameters using (22)–(23).

Table 1 Main parameters in the proposed model

Parameter	Symbol	Numerical value
Learning rate for actor	α_a	0.001
Learning rate for critic	α_b	0.002
Reward discount	γ	0.9
Soft replacement	tau	0.01
Memory capacity	D	10,000
Update batch size	m	32
Max episode	P	2500
Soft replacement rate	C	100
Max number of iterations	T	365
Epsilon greedy value	ε	0.01

$$\omega' \leftarrow \tau\omega + (1 - \tau)\omega' \tag{22}$$

$$\theta' \leftarrow \tau\theta + (1 - \tau)\theta' \tag{23}$$

(J) If S_{j+1} is the termination state, the current round iteration is completed, otherwise go to step (B).

The relevant parameters of the model are as Table 1.

5 Case Study

In this case, a typical VPP aggregated with various EVs is considered. Based on the training data of actual loads of DERs for 166 days [21], the benefits of the VPP under different pricing strategies are analyzed. The load of each EV is set to be 3 kW, and the power consumption for DR will shift during a day. The fixed service fee charging from DERs to a VPP is 0.5 yuan/(kWh), and the pricing range of the service fee is set from 0 to 1 yuan/(kWh). When the time-of-use electricity pricing is implemented, the period of electricity purchased from the grid is divided into three parts. The peak period is 12:00–21:00 with a peak price of 1.29 yuan/(kWh). The normal period is 7:00–12:00 and 21:00–24:00 with a normal price of 0.87 yuan/(kWh). The valley time is 0:00–7:00 with a valley price of 0.46 yuan/(kWh). The trading frequency of spot market and DR market is 15 min where DR market is open from 0:45 to 7:00 and from 12:45 to 21:00.

To verify the effectiveness and utility of the above methods, the DDPG algorithm simulates the settlement price in the DR market and the spot market, respectively. During the simulation, the training period was 166 days, and the daily service fee is formulated according to three pricing strategies, namely peak-valley time-of-use tariffs, hour pricing and quarter-hourly pricing. Comparing the income and load

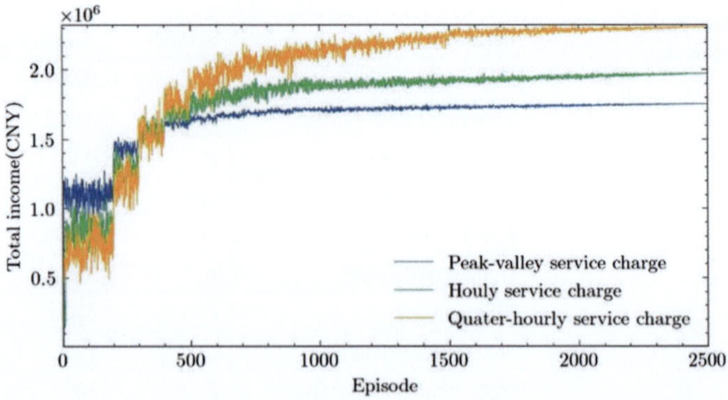

Fig. 4 Convergence of the algorithm and VPP's total incomes of the three scenarios

changes after the end of each cycle to obtain the effectiveness evaluation results of the strategies.

From the index dimension, the case compares the convergence degree of the algorithm under the three strategies and the VPP income under the variable prices. Evaluation metrics are obtained from application effect of the 166-day cycle run of the the peak-valley time-of-use tariff, hourly pricing, and quarter-hourly pricing strategy. The results are vividly presented in Fig. 4.

It can be seen that the overall benefit of a VPP under all three strategies has improved and stabilized at a higher level with increasing time. Due to the low output dimension of the algorithm, the output of the peak-valley time-of-use tariff strategy shows that a rapid gain increase is achieved at the initial round and the model converges rapidly around 500 iterations. The output result of the quarter-hourly pricing strategy rises in multiple rounds of volatility, and the return range is larger, showing that the initial round revenue is at a low level. The model gradually converges around 1500 iterations after experiencing a large periodic sharp rise. The output of hourly pricing strategy is somewhere between the two strategies. Overall comparison, the implementation of the quarter-hourly pricing strategy yields the highest additional income (2.31 million yuan), which is 1.18 times of the hourly pricing strategy and 1.31 times of the peak-valley time-of-use tariff.

When the total electricity consumption remains at a certain level, the price distribution and the average value of the three pricing strategies are relatively similar. In the peak and valley period in the DR market, the price output by the DDPG algorithm is close to the critical value. When in normal period without trading in DR market, the average valley price under the quarter-hourly pricing strategy is relatively low and had significant price fluctuations. The average daily service fee distribution of the quarter-hourly pricing strategy is shown in Fig. 5.

Figure 6 is a variable price fluctuation graph over time variation for the quarter-hourly pricing strategy. In the peak and valley period, the price fluctuation range is limited and the fluctuation trend is weak. In the normal period, the price fluctuation range is large and the fluctuation trend is obvious.

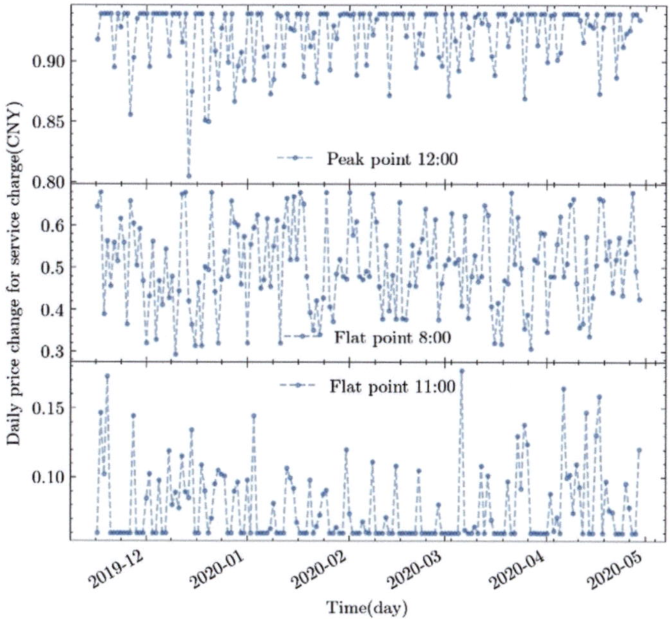

Fig. 5 The distribution of the average daily service charge under the quarter-hourly pricing

Fig. 6 Daily price fluctuation at typical time points of quarter-hourly pricing

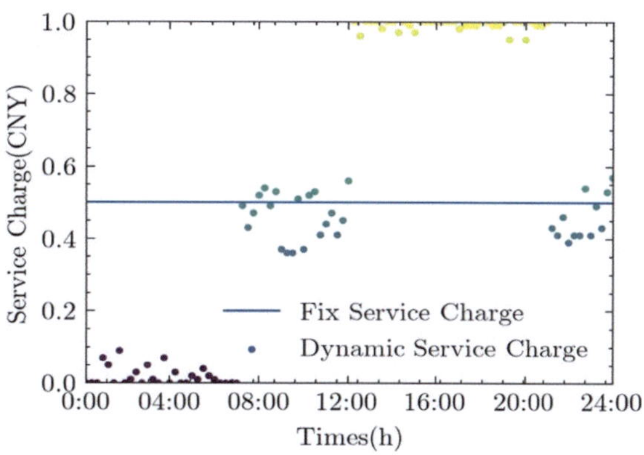

The average and standard deviation results of the three pricing strategies in different time periods are listed in Table 2. It can be intuitively seen that by increasing the pricing frequency, the price volatility of the standard deviation measurement increases significantly. The volatility in the flat-time period is higher than that in other periods, especially in the quarter-hourly pricing, with the normal period was about three times than that of the peak-valley period. Focusing on the price average vertically, the price average is almost equal at different pricing frequencies. The

Table 2 The mean and standard deviation of the three pricing strategies in different time periods

Pricing strategy		Peak-time period	Flat-time period	Valley-time period
Peak-valley pricing	u	0.976	0.498	0.036
	σ	0	0	0
Hourly pricing	u	0.973	0.475	0.032
	σ	0.016	0.029	0.012
Quarter-hourly pricing	u	0.982	0.461	0.022
	σ	0.024	0.069	0.026

average price of the normal period decreases slightly with the increase of pricing frequency.

The quarter-hourly pricing strategy has the characteristics of continuously tracking the external market prices, which is also the biggest advantage of this pricing model. To verify the physical characteristics of the pricing model under DDPG algorithm, the load-shifting effect is simulated and demonstrated in Fig. 7 and Table 3. The quarter-hourly pricing strategy performs better than the other strategies, and the peak load is significantly reduced. Among the reduced loads of DERs, 34.28% of the loads is divided into normal periods, while the corresponding result of hourly pricing is 28.72% and the result of peak-valley time-of-use tariff is 24.17%. The load-shifting effect during peak hours benefits from the flexible pricing of the dynamic pricing strategy.

The VPP revenue results of the three pricing methods are shown in Fig. 8. At the peak of around 12:00 to 21:00, the returns of the three strategies are significantly

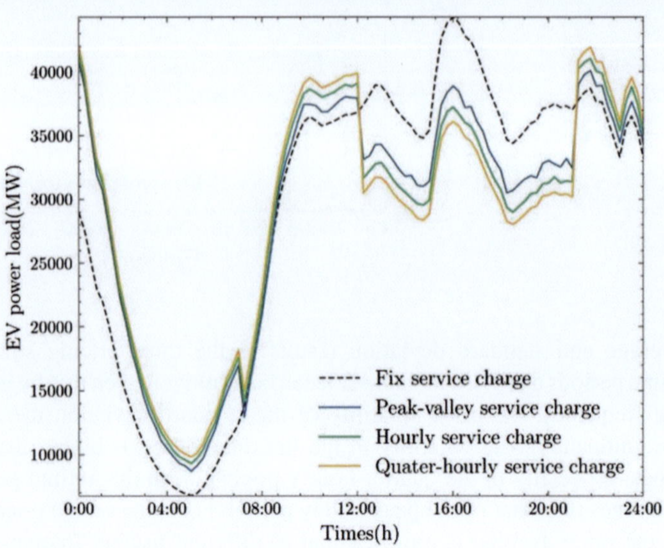

Fig. 7 Load variations under dynamic pricing

Table 3 Load variation range under dynamic pricing

Pricing strategy	Load change (MW)		
	Peak-time period	Flat-time period	Valley-time period
Peak-valley pricing	−172,581.43	33,596.73	138,984.70
Hourly pricing	−215,444.87	61,882.66	153,562.21
Quarter-hourly pricing	−254,062.48	87,091.26	166,971.22

larger than the fixed pricing strategy, but the revenue of the other periods is lower than the fixed pricing strategy. Among them, the revenue situation of the dynamic pricing strategy is optimal. Low revenue except peak time is due to the impact of the spot market, and price changes cause partial load shifting during peak periods.

Under the constraint of constant total electricity consumption, the incremental revenue mainly comes from the transaction revenue of the spot market and the DR market. The trading results are provided in Fig. 9 and Table 4. Figure 9a shows the comprehensive revenue increase of a VPP. Figure 9b shows the incremental revenue of the DR market, which is the main source of total revenue. Figure 9c shows the incremental revenue in the spot market. The spot market revenue is realized through the spread of market prices after the load shifting, and the net return is less than 1/10 of the total revenue under the constraint of the constant total load, despite the amount is about 2–3 times of the DR market. The DR market revenue is characterized by peak revenue which is lower than that of the valley period, with a ratio of about 1:2. Revenue increases significantly as the pricing frequency increases. Implementation

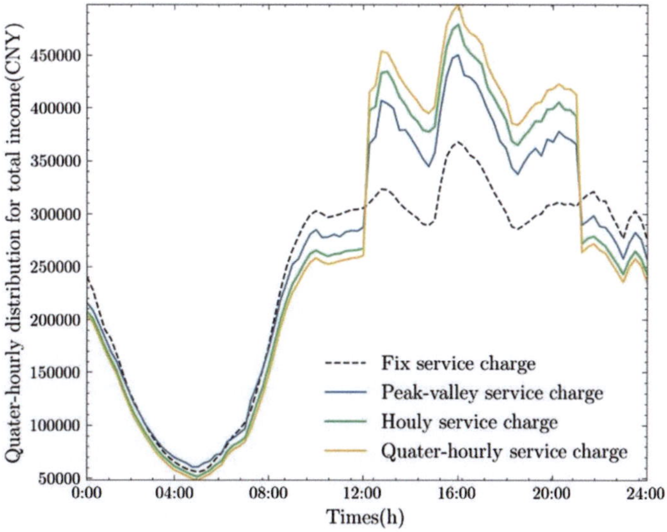

Fig. 8 Trend change of average return at time point

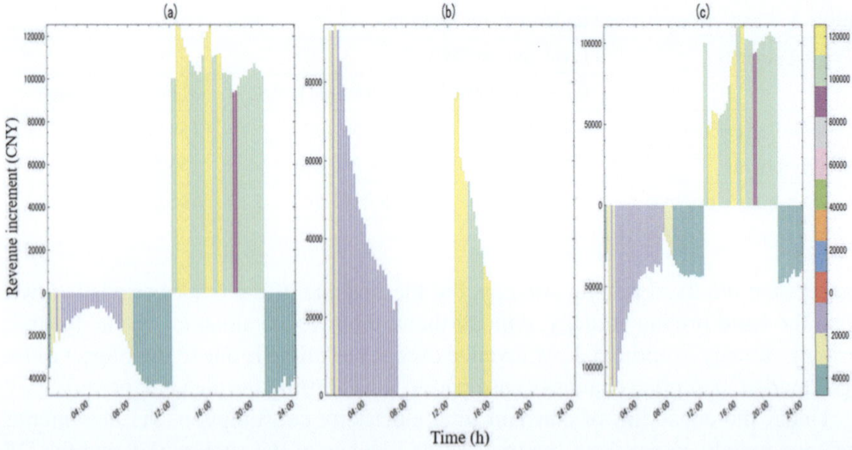

Fig. 9 Sources of incremental revenue

Table 4 Results of incremental income under different pricing strategies

Pricing strategy		Income increment (CNY)			
		Peak-time period	Flat-time period	Valley-time period	Total
Peak-valley pricing	R^s	2,211,088.5	−653,214.9	−1,450,829.7	107,043.8
	R^a	517,139.1	0	1,102,659.1	1,619,798.0
Hourly pricing	R^s	2,854,741.6	−1,047,224.8	−1,645,774.7	161,742.0
	R^a	596,699.9	0	1,241,207.2	1,837,907.2
Quarter-hourly pricing	R^s	3,368,595.1	−1,350,961.8	−1,826,777.7	190,855.6
	R^a	733,650.1	0	1,435,080.4	2,168,730.5

of the optimal dynamic (quarter-hourly) pricing strategy will increase about 10% of overall revenue for a VPP.

6 Conclusion

In order to fully adapt to the flexibility characteristics of DERs and VPPs, this chapter aims to maximize the revenue of VPPs participating in the spot market and DR market, and proposes dynamic pricing strategies based on variable market prices, which stimulates the scheduling potential and economic potential of DERs. The proposed strategy solves the discrete problem of the traditional pricing model for DERs through the output of the DDPG reinforcement learning algorithm.

The case study of algorithm comparison verifies the effectiveness of DDPG algorithm and output quotation. In a word, the higher the pricing frequency, the easier it is to form an updated market price quickly, so as to promote behavior of DERs to make expected changes and realize price guidance. The service price during normal time period is the key to creating different gains. While the pricing in both the valley period and the peak period tends to the limit, the average price fluctuates significantly, which is enough to cause VPPs to make response adjustment actions, shifting 18.5% of the peak load to the normal period and the valley period to generate large incremental revenue for DERs. Furthermore, the flexible scheduling potential of DERs allows VPPs to generate incremental revenue from spot market arbitrage. If effective controlling can make the energy cost of DERs remain at the same level, the VPP's overall operating revenue increases by 10%. Under the 15-min dynamic pricing strategy, achieving a significant 18% increase in earnings per hour is mathematical, while the peak load shifting volume increases 19% comparing to the hourly pricing strategy, and 47% comparing to the peak-valley pricing strategy.

References

1. D.F. Recalde Melo, A. Trippe, H.B. Gooi, T. Massier, Robust electric vehicle aggregation for ancillary service provision considering battery aging. IEEE Trans. Smart Grid **9**(3), 1728–1738 (2018)
2. D. Toquica, P.M. De Oliveira-De Jesus, A.I. Cadena, Power market equilibrium considering an EV storage aggregator exposed to marginal prices—a bilevel optimization approach. J. Energy Stor. **28**(3), 101267 (2020)
3. S.M. Bagher Sadati, J. Moshtagh, M. Shafie-khah, A. Rastgou, J.P.S. Catalão, Operational scheduling of a smart distribution system considering electric vehicles parking lot: a bi-level approach. Int. J. Electr. Power Energy Syst. **105**(1), 159–178 (2019)
4. S.I. Vagropoulos, D.K. Kyriazidis, A.G. Bakirtzis, Real-time charging management framework for electric vehicle aggregators in a market environment. IEEE Trans. Smart Grid **7**(2), 948–957 (2016)
5. C.D. Korkas, S. Baldi, S. Yuan, E.B. Kosmatopoulos, An adaptive learning-based approach for nearly optimal dynamic charging of electric vehicle fleets. IEEE Trans. Intell. Transp. Syst. **19**(7), 2066–2075 (2018)
6. H. Yi, Q. Lin, M. Chen, Balancing cost and dissatisfaction in online EV charging under real-time pricing. IEEE Conf. Comp. Commun., 1801–1809 (2019). https://doi.org/10.1109/INFOCOM.2019.8737558
7. Z. Liu, Q. Wu, K. Ma, M. Shahidehpour, Y. Xue, S. Huang, Two-stage optimal scheduling of electric vehicle charging based on transactive control. IEEE Trans. Smart Grid **10**(3), 2948–2958 (2019)
8. B. Hashemi, M. Shahabi, P. Teimourzadeh-Baboli, Stochastic-based optimal charging strategy for plug-in electric vehicles aggregator under incentive and regulatory policies of DSO. IEEE Trans. Veh. Technol. **68**(4), 3234–3245 (2019)
9. G.R. Aghajani, H.A. Shayanfar, H. Shayeghi, Demand side management in a smart micro-grid in the presence of renewable generation and demand response. Energy **126**(C), 622–637
10. S.M.B. Sadati, J. Moshtagh, M. Shafie-khah, J.P.S. Catalão, Smart distribution system operational scheduling considering electric vehicle parking lot and demand response programs. Electr. Power Syst. Res. **160**, 404–418 (2018)
11. L. Gong, W. Cao, K. Liu, J. Zhao, Optimal charging strategy for electric vehicles in residential charging station under dynamic spike pricing policy. Sustain. Cities Soc. **63**, 102474 (2020)
12. C.M. García Mazo, Y. Olaya, S. Botero Botero, Investment in renewable energy considering game theory and wind-hydro diversification. Energy Strategy Rev. **28**, 100447 (2020)

13. Z. Liu, Q. Wu, S. Huang, L. Wang, M. Shahidehpour, Y. Xue, Optimal day-ahead charging scheduling of electric vehicles through an aggregative game model. IEEE Trans. Smart Grid **9**(5), 5173–5184 (2018)
14. A. Tiguercha, A.A. Ladjici, M. Boudour, Competitive co-evolutionary approach to stochastic modeling in deregulated electricity market. IEEE Int. Energy Conf. (ENERGYCON) **2014**, 514–519 (2014)
15. A. Tiguercha, A.A. Ladjici, M. Boudour, Suppliers' optimal biding strategies in day-ahead electricity market using competitive coevolutionary algorithms. In *3rd International Conference on Systems and Control* (2013), pp. 821–826
16. J. Vijaya Kumar, D.M. Vinod Kumar, K. Edukondalu, Strategic bidding using fuzzy adaptive gravitational search algorithm in a pool based electricity market. Appl. Soft Comput. **13**(5), 2445–2455 (2013)
17. P. You, Z. Yang, M.-Y. Chow, Y. Sun, Optimal cooperative charging strategy for a smart charging station of electric vehicles. IEEE Trans. Power Syst. **31**(4), 2946–2956 (2016)
18. Z. Yan, Y. Xu, Data-driven load frequency control for stochastic power systems: a deep reinforcement learning method with continuous action search. IEEE Trans. Power Syst. **34**(2), 1653–1656 (2019)
19. R.S. Sutton, A.G. Barto, *Reinforcement Learning: An Introduction*, 2nd edn. (Bradford Book, 2018)
20. H. Dong, Z. Ding, S. Zhang, *Deep Reinforcement Learning: Fundamentals* (Springer, Research and Applications, 2020)
21. D. Liu, W. Wang, L. Wang, H. Jia, M. Shi, dynamic pricing strategy of electric vehicle aggregators based on DDPG reinforcement learning algorithm. IEEE Access **9**, 21556–21566 (2021)

Chapter 8
Reliable Operation of Power Systems Integrated with Virtual Power Plants

With information and communication technologies (ICT) developing and power systems deregulating, VPPs are able to participate in bidirectional interaction with the power grid. VPPs can provide operational reserves with flexible demand through load curtailment and shifting to assist in the operation of the power system. However, the reliability of the power system will be affected by several features such as load curtailment and shifting. In addition, customer engagement performance, random failures of ICS, and other complications due to the diversity of load types can also affect the system operation indirectly. In this chapter, the reliability network equivalent (RNE) technique will be applied in combination with time-sequential simulation (TSS) technique to study power system reliability. The RNE technique investigated the reserve capacity provided by VPPs, including chronological characteristics and uncertainties. This chapter also includes collaborative optimization of power generation and VPP deployment, and designs the optimal operation dispatch for system emergencies. The TSS method is used to evaluate the reliability of the power system. The modified IEEE RTS also validates the above method.

1 Introduction

Both renewable energy generation and demand-side resources are volatile and stochastic, requiring large operating reserves to maintain power system reliability. Operating reserve can be divided into three categories, which are primary reserve, secondary reserve and tertiary reserve, taking into account the time of the deployment after system disruptions [1]. Primary control is carried out within seconds of interference, the purpose of which is to keep the frequency of the system within a predetermined range with the help of a regulator, if there are small fluctuations in load or unexpected generating loss in the system. Secondary control takes place over several minutes, which is related to normal load monitoring, aiming at restoring frequency to nominal value and managing to load of transmission facilities [2]. The purpose of

H. Jia et al., *Business Models and Reliable Operation of Virtual Power Plants*,
https://doi.org/10.1007/978-981-19-7846-3_8

ensuring the tertiary reserve is to introduce a new post-contingency set point with zero area control error, which also complies with prescription limits to transmission flows [1]. The generation units provide a relatively stable operation reserve. However, the reliability and availability of power systems are still negatively affected by a lot of fluctuations of the time-varying load and the system emergencies [3]. Therefore, many scholars have studied the reliability evaluation of power systems from multiple practical scenarios [4–6], and used different methods in operation reserve arrangement, adaptive control, and operation reliability analysis, among which analysis methods such as decision-dependent uncertainty (DDU) [7], Markov process [8] and simulation technology [9] are widely used. Reliability network equivalent (RNE) technique is also utilized in analyzing restructured power systems, which can flexibly simulate power system characteristics and complicated operating scenarios, while also simplifies analysis procedures and workload [10–14].

Operational reserves provided by traditional generating units may be not economical and environment-friendly. The growing demand-side flexible loads (such as air conditioners and electric vehicles) can interact well with the grid to provide operational reserve as well [15]. By aggregating flexible loads in the demand side, VPPs can provide the power system with the required response services, such as tertiary reserves through load shifting and load curtailment. Load shifting means that the electricity customers reduce their electricity consumption during the peak period, while increasing the electricity consumption during the valley period, to obtain the corresponding compensation through the way of demand response. Load curtailment refers to the temporary suspension of some electricity load during or even contingency periods. In terms of electricity cost, by aggregating flexible resources in demand side to provide operating reserve, VPPs can reduce the system operating cost and carbon emissions, which are widely accepted and have a broad market prospect [16–18].

The intermittent nature of both solar power and wind power generation undoubtedly makes the effect of renewable energy generation on direct detriment to grid reliability. Similarly, the time-varying nature of the reserve capacity provided by VPPs based on load aggregation from load shifting and load curtailment also has significant impact on grid reliability. The stochastic nature of customer consumption on the flexible load also directly affects the amount of load aggregation, i.e., the amount of operating reserves [15]. The load aggregation process and the implementation of the two strategies mentioned above may also result in unexpected situations in the information and communication system (ICS) [19], which may cause process interruptions and reduce the reserve capacity. Therefore, the requirements to restructure the operating reserve capacity of the VPPS should focus on time-varying factors and multiple uncertainties, where a comprehensive quantitative analysis is performed.

Some scholars have quantitatively evaluated the reliability of the devices or resources involved in load aggregation systems. Some studies have analyzed the reliability and availability of load aggregation (not affected by demand response) formed by energy storage systems [20, 21]. Some studies considered electric vehicles as a flexible energy storage system and evaluated the impact caused by this system on the reliability of the superior power system by considering the load variability under participation in demand response [22]. However, the regulation strategies mentioned in a large number of studies still do not take into account enough factors such as chronological characteristics, for example, when studying the multiple uncertainties in VPPs providing operating reserves, the uncertainty of participation performance is considered in the estimation of the elasticity of demand response, while adding this variable to the random variables of the elasticity estimation does not accurately reflect the stochastic electricity consumption behavior of customers. To study the stochasticity and the range of adjustable capacity of flexible resources in demand side, the electricity consumption characteristics of the electricity-using entities, such as battery characteristics and driving characteristics, should be studied [23, 24].

Based on the analysis in [13] where flexible reserve provider on demand side was considered, in this chapter, the reliability impact of the tertiary reserve on the power system through VPPs by load aggregation on the demand side is studied, and the chronological characteristics of the two main strategies of load curtailment and load shifting are focused on. Three of the uncertainty factors analyzed in this tertiary reserve are power consumption behavior, ICS random faults and load type. The extension of RNE technology is utilized to enhance the computational analysis capabilities of demand response simultaneously, including the reserve capabilities provided by VPPs through equivalent multi-state VPP (EMVPP), such as chronological characteristics and uncertainty analysis. For power systems in which VPPs participate, an optimal power flow model is proposed considering the equivalent for multi-state generations, random failures of transmission network and reserve capacities provided by VPPs.

2 Reliability Network Equivalent for VPPs

Demand-side flexible resources generally include the electrical equipment of industrial users, commercial users and residential users, such as air conditioners, electric vehicles, heat pumps and other equipment. VPPs aggregate these flexible resources to provide tertiary reserve and interact well with the complex-structured reconfiguration power systems.

The impact of VPPs on the reliability of power systems can be analyzed by using reliability network equivalent (RNE) technology. First, a VPP is equivalent to an EMVPP through this technique as illustrated in Fig. 1. Then, it considers the time variation of load shifting and load curtailment with the consideration of analysis of three uncertain factors, including customers' participation performances, random failures of ICS and different load types.

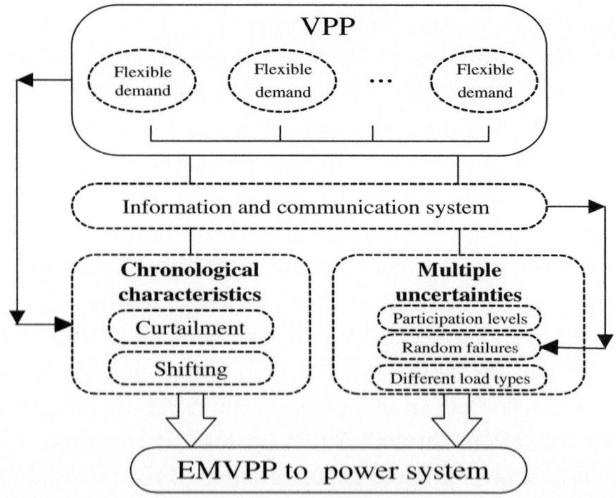

Fig. 1 RNE for VPPs

2.1 *Multi-state Model Considering Chronological Characteristics of VPPs*

In order to study the reserve capacity provided by VPPs, it is necessary to analyze the characteristics of VPPs. This chapter develops a conventional load model [25] while a multi-state model is utilized to formulate the time-changing characteristics of VPPs.

There is a system operator to set initial state, such as the duration of the study period T. For the T time period, the load can be divided into β sequence of load levels. Each set of sequence of load levels can be divided into M different load levels with different duration times D_l from the historical load data as shown in Fig. 2. Usually, β is equal to 1. The summation of the time duration of each load level is the total study period T as shown in Eq. (1). For example, the total study period T can be one week.

$$T = \beta \sum_{l=1}^{M} D_l \tag{1}$$

The presented multi-state load model in Reference [25] illustrates β sequence of M load levels which are sequentially connected as they appear in the historical load data. In the Markov load model, the transition rate of the lth load level λ_l can be obtained as Eq. (2).

$$\lambda_l = 1/D_l \tag{2}$$

There exist uncertainties in the system load as well as the variation with time. The load model is supposed to reflect not only the uncertainties in the load levels but also the durations of these load levels. Markov load model is obtained after collecting

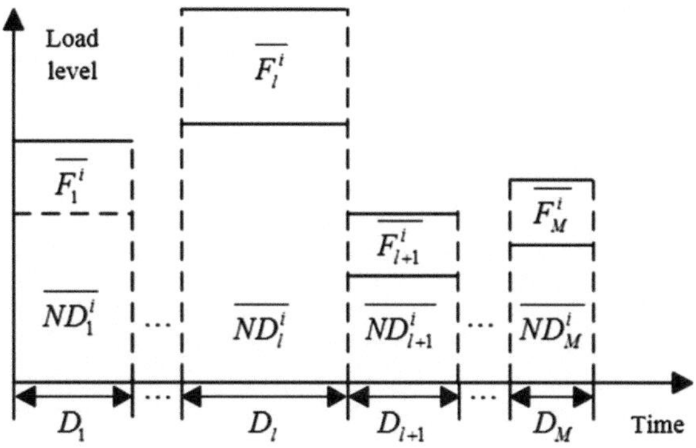

Fig. 2 The sequence of load levels

different load levels and load transiting rates at various levels, as shown in Fig. 3, which is a widely used reliability assessment method [26–28]. The key point of the Markov load model is to maintain the chronological representation and uncertainties in the durations of different load levels. To make the model more accurate, the number of sets of sequence of load levels in a study period T can be reduced to 1 where there is no repeat for sets of sequence. The number of groups at the load level should be as large as possible to make the model more convincing, as far as the experimental conditions allow.

The load can be divided into M states which are sequentially connected and located at N buses. $L_l^i(t)(i = 1, \ldots, N; l = 1, \ldots, M)$ indicates the load in the l th state at the i th bus. The multi-state load model imposes the same variation pattern for all load buses. Therefore, the assumption that all the bus loads are perfectly correlated are made [25].

The load can be divided into shiftable load, curtailable load, and non-dispatchable load. The shiftable load and curtailable load can be uniformed as flexible load. Especially, there exist different types of shiftable load. The capacity of shiftable load can be divided into m_l types at bus i for state l. The capacity of the m_{th} type

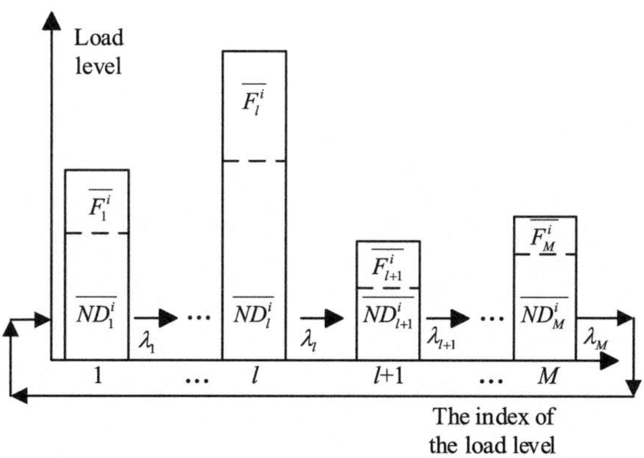

Fig. 3 Multi-state load model at bus i

of shiftable load at bus i for state l can be represented as $\overline{FS_l^{i,m}}(t)$. The capacities of different load types at bus i for state l can be denoted as $\overline{FS_l^i}(t)$, $\overline{FC_l^i}(t)$, and $\overline{ND_l^i}(t)$, where $\overline{FS_l^i}(t) = \sum_{m=1}^{m_l} \overline{FS_l^{i,m}}(t)$. The capacity of flexible load at bus i for state l is $\overline{F_l^i}(t)$. The proportions of each load type can be different and formulated as $\alpha_1, \alpha_2, \alpha_3$. Then, we have

$$\begin{cases} \overline{FS_l^i}(t) = \alpha_1 L_l^i(t) \\ \overline{FC_l^i}(t) = \alpha_2 L_l^i(t) \\ \overline{ND_l^i}(t) = \alpha_3 L_l^i(t) \\ \alpha_1 + \alpha_2 + \alpha_3 = 1 \end{cases} \tag{3}$$

Especially, for the shiftable load, different types of shiftable load at bus i for state l can be represented as a vector $\overline{\boldsymbol{FS}_l^i}(t)$.

$$\overline{\boldsymbol{FS}_l^i}(t) = \left[\overline{FS_l^{i,1}}(t), \ldots, \overline{FS_l^{i,m}}(t), \ldots, \overline{FS_l^{i,m_l}}(t)\right]^T \tag{4}$$

The maximum shiftable time period of the m_l types of shiftable load can be demonstrated as a vector $\boldsymbol{d}_l^i$:

$$\boldsymbol{d}_l^i = \left[d_l^{i,1}, \ldots, d_l^{i,m}, \ldots, d_l^{i,m_l}\right] \tag{5}$$

The reserve capacity provided by VPP from different types of flexible load for bus i at time t can be formulated as a vector $\boldsymbol{OR}_l^i$:

$$\boldsymbol{OR}_l^i = \begin{bmatrix} \overline{\boldsymbol{FS}_l^i}(t) \\ \overline{FC_l^i}(t) \end{bmatrix} \tag{6}$$

Figure 4 illustrates the multi-state load model when curtailment and shifting happen based on the conventional load model. The original flexible load capacity can be represented as:

$$\begin{aligned} \overline{F_l^i}(t) &= FS_l^i(t) + FC_l^i(t) + FU_l^i(t) \\ &= \sum_{m=1}^{m_l} FS_l^{i,m}(t) + FC_l^i(t) + FU_l^i(t) \end{aligned} \tag{7}$$

In Eq. (7), $FS_l^i(t)$, $FC_l^i(t)$ and $FU_l^i(t)$ indicate the deployed shifted load, deployed curtailed load, and unused capacity of VPP for state l at bus i, respectively. $FS_l^{i,m}(t)$ denotes the deployed shifted load of the m_{th} type shiftable load. The

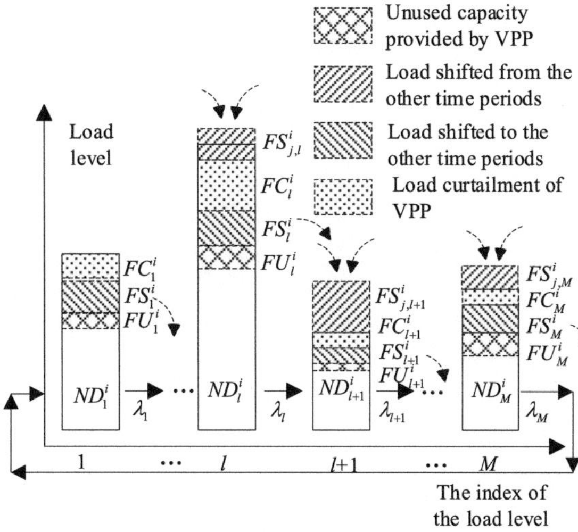

Fig. 4 Multi-state load model when curtailment and shifting happen

vector $\boldsymbol{FS}_l^i(t)$ can be utilized to present the deployed shifted load from the m_{th} type of shiftable load at bus i.

$$\boldsymbol{FS}_l^i(t) = \left[FS_l^{i,1}(t), \ldots, FS_l^{i,m}(t), \ldots, FS_l^{i,m_l}(t)\right]^T \tag{8}$$

The load shifted from the lth state to the jth state can be represented as $\boldsymbol{FS}_{l,j}^i(t)(j \neq l)$, where

$$\boldsymbol{FS}_{l,j}^i(t) = \left[FS_{l,j}^{i,1}(t), \ldots, FS_{l,j}^{i,m}(t), \ldots, FS_{l,j}^{i,m_l}(t)\right]^T \tag{9}$$

$FS_{l,j}^{i,m}(t)(j \neq l)$ is the deployed load shifted from the lth state to the jth state.

It should be noted that if the shiftable time period $d_l^{i,m}$ is smaller than the time interval between the lth state and the jth state $\Delta t_{l,j}$, the mth type shiftable load would not be deployed. It can be formulated as Eq. (10).

$$FS_{l,j}^{i,m}(t) = 0, if\ d_l^{i,m} < \Delta t_{l,j} \tag{10}$$

Moreover, considering the possible different durations of time periods, the constraint that the deferred energy from load shifting equals to the incurred energy is conducted. In the proposed model, the shifted energy from the lth load level is equal to the energy shifted to the other load levels. That is to say, for the lth state, the shifting energy should meet Eq. (11).

$$\boldsymbol{FS}_l^i(t) \cdot t_l = \sum_{j \neq l}^{M} \boldsymbol{FS}_{l,j}^i(t) \cdot t_j \tag{11}$$

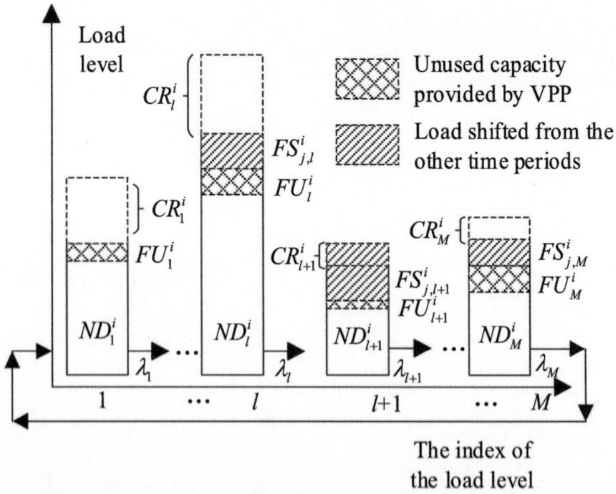

Fig. 5 Multi-state load model after curtailment and shifting

where t_l indicates the duration time of the lth load level in the Markov load model.

For the load curtailment and load shifting, the corresponding strategy is to minimize the total system cost utilizing optimal power flow technique. More details are demonstrated in Sect. 3.2.

After the curtailment and shifting processes of VPPs, the load at bus i can be updated as presented in Fig. 5.

The updated load $L_l^{i'}(t)$ in the lth state at bus i can be formulated as:

$$\begin{aligned} L_l^{i'}(t) &= ND_l^i(t) + FU_l^i(t) + \sum_{j\neq l}^{M}\sum_{m=1}^{m_l} FS_{j,l}^{i,m}(t)(l = 1,\ldots,M) \\ &= L_l^i(t) + \sum_{j\neq l}^{M}\sum_{m=1}^{m_l} FS_{j,l}^{i,m}(t) - FC_l^i(t) - FS_l^i(t)(l = 1,\ldots,M) \\ &= L_l^i(t) + \sum_{j\neq l}^{M}\sum_{m=1}^{m_l} FS_{j,l}^{i,m}(t) - FC_l^i(t) - \sum_{j\neq l}^{M}\sum_{m=1}^{m_l} FS_{l,j}^{i,m}(t)(l = 1,\ldots,M) \end{aligned} \tag{12}$$

The updated load of the system is:

$$L_l'(t) = \sum_{i=1}^{N} L_l^{i'}(t) \tag{13}$$

For the bus i, the actual contribution of load shifting and load curtailment from VPPs in the lth state can be formulated as $CR_l^i(t)$:

$$CR_l^i(t) = L_l^{i'}(t) - L_l^i(t) = \sum_{j\neq l}^{M}\sum_{m=1}^{m_l} FS_{j,l}^{i,m}(t)$$
$$- FC_l^i(t) - \sum_{j\neq l}^{M}\sum_{m=1}^{m_l} FS_{l,j}^{i,m}(t)\ (l = 1, \ldots, M) \quad (14)$$

From the overall prospective, for the lth state at bus i, if $CR_l^i(t) > 0$, it indicates the load increases after the adjustment of demand response and VPPs provide down reserve for the system. Conversely, the load reduces and VPPs provide up reserve for the system.

The total contribution of load shifting and load curtailment from VPPs to the system can be calculated by the summation of the actual contribution for each bus:

$$CR_l(t) = \sum_{i=1}^{N} CR_l^i(t) \quad (15)$$

2.2 Multi-state Model Considering Uncertainties from VPPs

The operational reserve capacity of VPPs is associated with the uncertainties from VPPs, besides influencing by chronological factors. These uncertainties from VPPs arise from the customer engagement performance, the random failure of the ICS and various types of load. A multi-state model can well describe the uncertainties of VPPs due to the above factors. For example, the state of user participation in a VPP, from no participation at all (0% proportion) to full participation (100% proportion) setting different participation percentage levels to quantify the uncertainty from user behaviour. Multi-state model considering uncertainties of customers' participation levels and random failures of ICS are shown in Fig. 6.

The participation level of VPPs is dependent on the behaviors of customers to some extent and can be derived from historical database. Customers' electricity consumption behaviors may be effected by factors like the incentive mechanism or

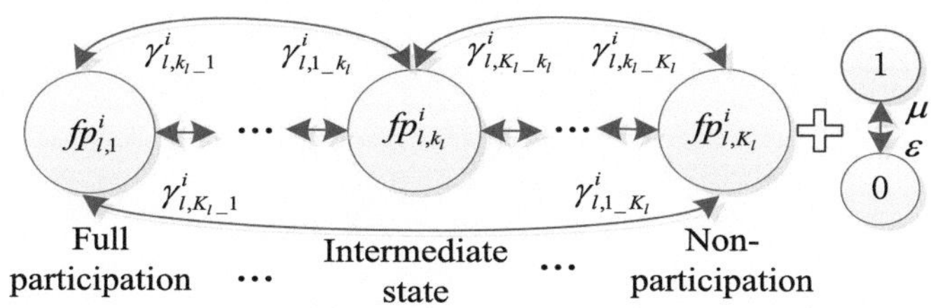

Fig. 6 Multi-state model considering uncertainties of customers' participation levels and random failures of ICS

weather conditions. fp^i_{l,k_l} represents the participation level of VPPs. There are K_l participation levels for the l_{th} state. Specifically, $fp^i_{l,1}$ and fp^i_{l,K_l} represent VPPs in the full participation state and non-participation state, respectively. The transition rate of participation level from a to b is designated as γ^i_{l,a_b}.

Remarkably, the development of information and communication technologies has made significant contributions to power systems. ICS has become an indispensable part to realize the active participation of VPPs under the smart grid framework. However, uncertainties from ICS including random failures of its physical system may affect the participation level of VPPs. State-space based methods have been developed to capture necessary information about the physical system [29]. Binary-state models have been widely applied in reliability evaluation for engineering systems [30] to demonstrate the perfect functioning state and complete failure state. In this part, we use binary-state model to illustrate the reliability model of uncertainties from ICS for VPPs. The failure rate and repair rate of ICS are presented as ε and μ. If the ICS is in perfect functioning state, the state is represented as "1". Otherwise, if the ICS is in complete failure state, the state is represented as "0".

Moreover, it should be noted that the failure of ICS would impact the implementation of load shifting as presented in Fig. 7. If the ICS is in failure in the lth time period after it starts t_l^{fail} time, the load shifting would be stopped immediately. In this case, the mth type of shiftable load shifted from other time periods to the lth time period $FS^{i,m}_{j,l}(j \neq l)$ would only last for t_l^{fail} time and cannot be fully recovered. Therefore, the shifted load with the remaining $(t_l - t_l^{fail})$ time in the lth time period which is not recovered is supposed to be calculated in the reliability indices. The energy loss that should be supplemented in the reliability indices under this condition is presented as Eq. (16).

$$SI^i_l = \sum_{j \neq l}^{M} \sum_{m=1}^{m_l} FS^{i,m}_{j,l}(t) \cdot (t_l - t_l^{fail}) \tag{16}$$

In the model description above, we assume that the value of non-dispatchable load, curtailable load and shiftable load are predesigned without uncertainties. However, in reality, there exist uncertainties which are supposed to be considered. The uncertainties from proportions of different load types as presented in Eq. (3) can be reasonably

Fig. 7 The failure of ICS

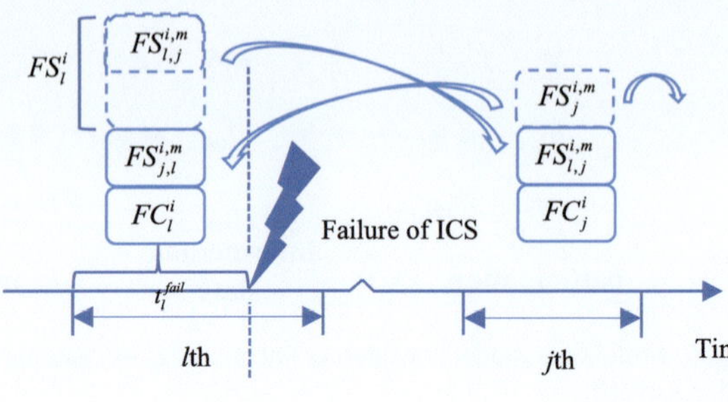

described by normal distributions $N \sim (\alpha_q, \sigma_q^2)$(q = 1,2,3). α_q and σ_q are the mean value and standard deviation of the proportion for the qth type of load.

2.3 *The EMVPP Considering Both Chronological Characteristics and Uncertainties*

The multi-state model of the EMVPP is built on the consideration of chronological characteristics and uncertainties of VPPs as illustrated in Fig. 8, where VPPs can arrange reserve capacity through power pools or bilateral contracts to adapt to the current features of restructured power systems.

The available reserve capacity considering uncertainties provided by EMVPP at time t should be revised based on Eq. (6) and represented as Eq. (17):

$$\boldsymbol{AR}_l^i = \boldsymbol{OR}_l^i \cdot fp_{l,k_l}^i = \left[\frac{\overline{\boldsymbol{FS}_l^i}(t)}{FC_l^i(t)} \right] \cdot fp_{l,k_l}^i \tag{17}$$

which takes values from $\boldsymbol{OR}_l^i \cdot \{fp_{l,1}^i, \cdots, fp_{l,k_l}^i, \cdots, fp_{l,K_l}^i\}$.

The TSS procedures to determine the characteristics of the EMVPP are elaborated as follows.

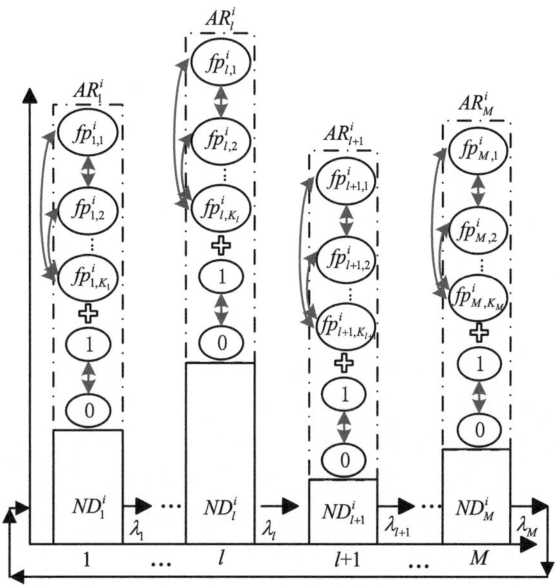

Fig. 8 Multi-state model for EMVPP

Step 1: Simulate the multiple state sequences of original load including the state and corresponding state duration time which follows the corresponding exponential distribution as shown in Sect. 2.1.

The duration sequence of each load state is determined by the load state transition rates. The load state transition rates λ_l are assumed to be known based on the historical load data. For the simulation, the duration time of the load state can be obtained as $t_l = -\frac{1}{\lambda_{l,l+1}} \ln U$, where U is a uniformly distributed random variable over the interval (0, 1) [31]. Simulating the load sequence utilizing this technique until the whole study period T is satisfied.

Step 2: Suppose the proportion of the capacity for different types of flexible load, including the curtailable load and different types of shiftable load in the operation period based on the original load sequence.

Step 3: For each state l, simulate the state sequences of participation level of VPPs which follow the corresponding exponential distribution in Sect. 2.2 utilizing the similar technique in Step 1.

For state l, the duration sequence of participation level of VPPs is determined by the transition rates γ^i_{l,a_b} between participation levels. Considering the transition rates of customer's participation from state a to state b for the l_{th} load level at bus i, the duration time for the ath participation level is $t^i_{l,a\_b} = -\frac{1}{\gamma^i_{l,a\_b}} \ln U (b \neq a)$, where U is a uniformly distributed random variable over the interval (0,1) [31]. Then, find the minimal duration time $t^{i,\min}_{l,a\_b}$ from all t^i_{l,a_b} where $b \neq a$. The ath participation level will transit to the bth participation level with the minimal duration time $t^{i,\min}_{l,a\_b}$. Since the transition rates between different participation levels are formally dependent on the load state l, utilizing this approach, the participation level can be obtained for each state l. Repeat this process for all load states, the sequences of participation level can be obtained for the whole study period.

Step 4: Simulate the state sequence of ICS based on the failure rate and repair rate utilizing the similar technique in Step 3.

Step 5: Determine the actual participation level utilizing the participation level of VPPs multiplied by the state of ICS.

Step 6: Make the reserve capacities of VPPs multiplied by the actual participation level to determine the available reserve capacities provided by VPP according to Eq. (17).

Step 7: Formulate the optimal power flow in Sect. 3 to determine the deployed reserve provided by VPPs. In the optimization problem, the load shifting, load curtailment and customer interruption are conducted.

3 Optimal Operation Dispatch for Contingencies Considering VPPs

When a power system is in peak hours or in contingency situations, aggregation and power dispatching in EMVPP will change. Generation and reserve capacity need to

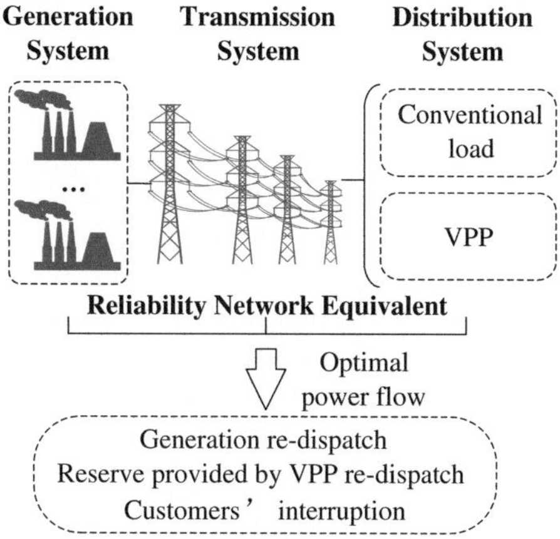

Fig. 9 Optimal operation dispatch for system in contingency state

be adjusted, and load interruptions occur for part of power consumers to ensure the safety and stability of the whole system.

The whole restructured power system basically includes the generation system, transmission system and loads which can be aggregated as VPPs. Therefore, the reliability of the overall system will be considered separately. First, through the RNE technique, the stochastic nature of the generation system is discussed to determine the impact of the generation capacity. Then, for the transmission system, an operation dispatch based on optimal power flow is designed to take transmission system failures into consideration. Finally, the overall system is optimized to pursue the minimum system cost, and Fig. 9 below graphically demonstrates the analysis process of the above optimal operation.

3.1 RNE for Generation System

The generation system consisting of conventional generating units is often used as a primary operating reserve, which is possible in both the long-term power-trading market and in the real-time market [32], and remains the same functions in complex restructured power systems. This means that conventional generating units can function in an economic dispatch state and provide operating reserve in an emergency state, which is an important support for the reliability of power systems.

Through the RNE technique, the multi-state generation system (EMGS) replaces the simple binary state and is used in this chapter, which takes full account of the controllable and stochastic nature of conventional generating units. Through the TSS technique, the characteristics of the EMGS are further analyzed.

3.2 Operation Dispatch Based on Optimal Power Flow

The operational dispatch is designed based on an optimal power flow model to dispatch generation and the operating capacity of VPPs at minimum cost, taking stochastic failures of the transmission network and the cost-effectiveness of restructured power systems into account. The stochastic failure rates and maintenance rates (or repair rates) of system components are important variables, and these data are usually collected from historical state data and preliminary analysis [33]. Having obtained the above model and parameters, the next step is to design reasonable scenarios and operate accordingly. Operators and dispatchers familiar with actual operating scenarios will be helpful, considering the chronological characteristics of the load, the duration of the dispatcher's decision on a particular load state may be much shorter, compared to the study period.

Within the above design framework, the objective function of the model is to minimize the system cost to meet the cost-effectiveness of restructured power systems. The system cost consists of four components, namely generation cost, reserve cost of generating units, compensation cost for VPPs through load curtailment and load shifting, and load interruption cost for electricity consumers. Since the load level l is dependent with time t, t is omitted for simplification in the optimization problem.

$$\begin{aligned}\text{Min}\quad TC = \sum_{l}^{M}\sum_{i=1}^{N}\Bigg\{&\sum_{ge=1}^{NG} GC_{l,ge}^{i}(P_{l,ge}^{i}) + \sum_{gr=1}^{NR} RC_{l,gr}^{i}(P_{l,gr}^{i})\\ &+ \sum_{c=1}^{NC} CC_{l,c}^{i}(FC_{l}^{i}) + \sum_{s=1}^{NS}\sum_{m=1}^{m_l} SC_{l,s}^{i,m}(FS_{l}^{i,m}) + \sum_{h=1}^{NH} IC_{l,h}^{i}(LI_{l}^{i})\Bigg\} \qquad (18)\end{aligned}$$

Generally, the electricity utilities and the service providers enter into a transaction through the ancillary services market. The common form of demand response stipulates that the service provider is required to carry out load shifting and load curtailment at an agreed time, with the actual demand response time being assessed and compensation paid at the end of the service. An interruption to an electricity customer's load is an outage, which for commercial and industrial customers means stopping production, resulting in economic losses [34]. Therefore, considering load interruptions as a reliability indicator, interruption costs are calculated with reference to the customer types and duration of the interruption, so as to figure out the customer damage functions for different types of electricity customers. The US Department of Energy conducted a study of load interruption costs for different types of electricity customers. The expected cost of unserved energy, which is the most widely used load interruption cost indicator, is utilized in this chapter.

Equation (18) is subject to the following constraints:

Power balance constraints:

$$(\boldsymbol{B} \cdot \boldsymbol{\theta}_l)^i = \sum_{ge=1}^{NG} P_{l,ge}^i + \sum_{gr=1}^{NG} P_{l,gr}^i - L_l^{i'} \tag{19}$$

Generation limits:

$$\underline{P_{l,ge}^i} \leq P_{l,ge}^i \leq \overline{P_{l,ge}^i} \tag{20}$$

Limits of reserve capacity from generation:

$$\underline{P_{l,gr}^i} \leq P_{l,gr}^i \leq \overline{P_{l,gr}^i} \tag{21}$$

Capacity limits of VPPs considering both curtailment and shifting:

$$\mathbf{0} \leq \begin{bmatrix} \boldsymbol{FS}_l^i \\ FC_l^i \end{bmatrix} \leq \boldsymbol{AR}_l^i \tag{22}$$

Customer interruption constraints:

$$0 \leq LI_l^i \leq \overline{LI_l^i} \tag{23}$$

Line flow constraints:

$$|\frac{1}{x_l^{i\_i'}}(\theta_l^i - \theta_l^{i'})| \leq |F_{\max}^{i\_i'}| \tag{24}$$

Constraints from load shifting as presented in Eqs. (10) and (11).

In the formulation of the optimization problem, $\boldsymbol{B}$ and $\boldsymbol{\theta}_l$ represent the admittance matrix of transmission network and the phase angle vector of bus voltages, respectively. $\overline{P_{l,ge}^i}$ and $\underline{P_{l,ge}^i}$, $\overline{P_{l,gr}^i}$ and $\underline{P_{l,gr}^i}$ are the upper limit and lower limit of power generation and generation reserve, respectively. $\overline{LI_l^i}$ is the upper limit of customer interruption. It is determined by the characteristic of the load. $x_l^{i\_i'}$ and $F_{\max}^{i\_i'}$ denote the reactance and maximum power flow of the transmission line between bus i and bus i', respectively.

It should be noted that if there is a failure of ICS in the system operation, customers are supposed to be interrupted. The total interruption cost TC' will be updated by supplementing the interruption cost caused by the failure of ICS as shown in Eq. (25). It should be noted that during the system operation, if a failure of ICS occurs, there might be unsuccessful response where shiftable load cannot be recovered due to the failure of ICS. Thus, the consequences including the cost and interrupted load [Eqs. (16) and (25)] are supposed to be modified when decisions on the system operation are implemented.

$$TC' = TC + \sum_{l=1}^{M}\sum_{h=1}^{NH} IC_{l,h}^{i}[\sum_{j\neq l}^{M}\sum_{m=1}^{m_l} FS_{j,l}^{i,m}(t)\cdot(t_l - t_l^{fail})] \quad (25)$$

4 TSS Procedures for Reliability Evaluation of Power Systems with VPPs

The suitable selection of reliability indicators is important for evaluating the reliability of power systems. In this chapter, the highly representative indicators, the expected energy not supplied (EENS) and instant loss of load probability (LOLP) are chosen to evaluate the restructured power systems in order to synthesize the multiple influencing factors mentioned earlier in this chapter. It is worth noting that the shifted load is not restored during the simulation due to ICS failure and needs to be additionally calculated to the reliability evaluation metrics in the following equations.

$$EENS(T) = \left\{\sum_{n=1}^{n_s}\sum_{i=1}^{N}\left[\int_0^T LI_l^i(t)\cdot dt + \sum_{l=1}^{M} SI_l^i\right]\right\}/n_s \quad (26)$$

$$EENS^i(T) = \left\{\sum_{n=1}^{n_s}\left[\int_0^T LI_l^i(t)\cdot dt + \sum_{l=1}^{M} SI_l^i\right]\right\}/n_s \quad (27)$$

$$LOLP(t) = \sum_{j=1}^{n_s} m_j(t)/n_s \quad (28)$$

where n_s indicates the sampling number of the simulation. In Eq. (28), $m_j(t) \in \{0, 1\}$. If there is no load interruption and non-recovered shifted load, $m_j(t) = 0$; Otherwise, if there is load interruption or non-recovered shifted load, $m_j(t) = 1$.

The steps below are the TSS procedures for evaluating reliability of the restructured power systems considering VPPs.

Step 1: Generate the state sequence for the EMGS at each bus for the study time period, e.g., one week, utilizing the TSS technique. Determine the state of the EMGS at time t according to the sampled sequence.

Step 2: Generate the state sequence for the load at each bus for the study time period utilizing the procedures described in Sect. 2.3. Determine the state of the load at time t as well as the available reserve capacity of VPPs according to the sampled sequence.

Step 3: The optimal power flow model introduced in Sect. 3.2 is utilized to assess the customer interruption at each bus at time t. If the system suffers from customer interruption, then the system is in failure state at time t.

Step 4: According to the state sequence of ICS, determine the shifted energy which is not recovered due to the failure of ICS based on Eq. (16).

Step 5: Go to Step 1 until the confidence intervals are satisfied. Otherwise, go to Step 6.

The stopping criterion given for the simulation is the EENS coefficient of variation.

$$\beta_{EENS} = \sqrt{V(EENS(T))}/EENS(T) \tag{29}$$

where V(EENS(T)) is the variance of EENS(T).

Step 6: Calculate the average reliability indices of the power system according to Eqs. (26)–(28) and update the total system cost if there is a failure of ICS.

5 Case Studies

The modified IEEE RTS validates the above method, which is restructured to become a whole system consisting of four 575-MW coal thermal generating units (utilized in real engineering systems), two 197-MW oil thermal generating units and five 50-MW hydro generating units [33]. The initial state of the various units can be "all-up" or not, because sometimes several units are in "failure" state, and they are supposed to be set by the system operator. It is studied that the generating units in "failure" state work as a variable or control group in the research on the reliability evaluation of power systems [35]. In this chapter, we set the whole system as all generating units working in the state with installed generation capacities respectively. The locations and bidding coefficients for the power generation [36] are presented in Table 1. The utilization cost for the generation reserve is assumed to be 20$/MWh higher than its generation price.

Table 1 The locations and unit costs for the generation units and VPPs

Generating unit	Location (bus)	Bidding coefficient	
		A ($/MW^{2}h)	B ($/MWh)
50 MW * 3	2	0.1	0.5
197 MW * 2	13	0.4	50.6
575 MW * 4	15, 16, 18, 23	0.05	12.1
50 MW * 2	21	0.1	0.5
VPPs	Location (bus)	Cost ($/MWh)	
		Shifting	Curtailment
Residential	1, 2, 3, 4, 5	50	100
Commercial	6, 7, 8, 9, 10, 13	70	150
Industrial	14, 15, 16, 18, 19, 20	70	150

Table 2 Transition rates of different load levels for weekdays and weekend

Transition rates (/h)	Valley	Flat 1	Peak	Flat 2
Valley	–	1/4, 1/6	0	0
Flat 1	0	–	1/3, 1/2	0
Peak	0	0	–	1/13, 1/12
Flat 2	1/4, 1/4	0	0	–

The study period for the above system is 168 h in a week. The load levels are with five weekdays and one weekend. The hourly peak load over a daily time period in a summer week presented in [37] can be divided into four categories: valley (2–6 h, 58%), flat 1 (6–9 h, 87%), peak (9–22 h, 100%) and flat 2 (22–2 h, 87%) in weekdays and valley (2–8 h, 66%), flat 1 (8–10 h, 86%), peak (10–22 h, 100%) and flat 2 (22–2 h, 88%) in weekend. The transition rates of different load levels for weekdays and weekend are shown in Table 2. The system operator determines the beginning time at certain point. Apparently, the initial state of the load is the load level at the beginning (at time 0 in a weekday) of the whole study period. It is a flat load level with 87% of the peak load.

The interruption of customers indicates a blackout and it could incur economic loss for customers. Therefore, it is essential to consider this kind of interruption into reliability indices. The interruption cost of customers (IC) depends on the types of customers, duration of the interruption, and so on [36]. Moreover, the costs of interruption for industrial and commercial customers are supposed to be higher than that for residential customers. Surveys have been conducted in Reference [38] to estimate the interruption cost of customers. The interruption costs of customers (IC) are formulated as customer damage functions for different types of customers, which are functions of the interruption durations [38]. Different customer damage functions are presented in Fig. 10 for different types of customers. Moreover, the curtailment and shifting of VPPs are under the framework of demand response. The compensation costs for curtailment and shifting can be determined by the contracts between the electricity utilities and customers or the bids of demand response customers in the energy market.

The compensation cost for the load shifting and load curtailment are presented in Table 1 as well. The interruption cost per unserved kWh for customers with different interruption duration [39] are presented in Table 3. The upper limit of customer interruption is the amount of total load which denotes all of the load can be interrupted when the system contingency occurs. The stopping criterion for the simulation is $\beta_{EENS} \leq 0.01$.

The main contents of the five case studies are as follows: Case 1 investigates the impact of VPP capacity variation, Case 2 investigates the impact of three internal factors of the VPP, Case 3 investigates the impact of shiftable time periods, Case 4 investigates the impact of the location (bus) of the VPP, and Case 5 investigates the load model with fixed time steps.

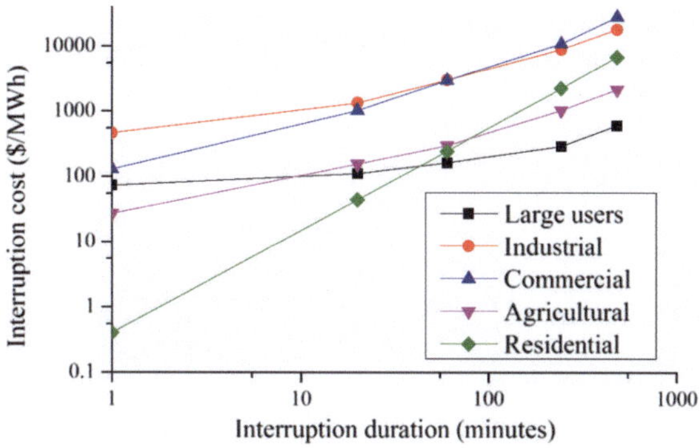

Fig. 10 Customer damage functions for different types of customers

Table 3 The interruption cost for different customers with different interruption durations

Cost ($/unserved kWh)	Interruption duration (h)				
	0.01	0.5	1	4	8
Residential	16.8	3.5	2.2	1.2	0.9
Commercial/industrial	96.5	22.6	15.3	13.0	10.6

5.1 *Case 1: Different Operating Reserve Capacities of VPPs*

In Case 1, four scenarios of the operating reserve are shown in Table 4, where VPPs provide with various proportions of load shifting and load curtailment. It should be noticed that the load change at fixed time periods within a day, and another assumption is no uncertainty from VPPs' ability control and availability.

Scenario A without load shifting or curtailment is the base scenario. During the one-week study period, the system EENS differences between Scenario B, Scenario C, Scenario D and Scenario A relatively are shown in Table 5. The time-varying LOLP is depicted in Fig. 11. The EENS and LOLP results, which reflect system reliability metrics, suggest that systems with more VPP capacity are more reliable than systems with less VPP capacity. The results of system costs shown in Table 6 demonstrate that, when the capacity of VPPs increases, i.e., the amount of regulation increases, the compensation cost of shifting and curtailment increases accordingly.

Table 4 The proportions of shiftable load and curtailable load for different scenarios in case 1

Scenario	A	B	C	D
Load shifting	0	0.05	0.1	0.15
Load curtailment	0	0.05	0.1	0.15
VPP	0	0.1	0.2	0.3

Table 5 System EENS and relative differences for scenarios in case 1

Scenario	EENS (MWh)	Relative difference
A	1142.51	0
B	395.47	−65.38%
C	101.38	−91.13%
D	63.43	−94.45%

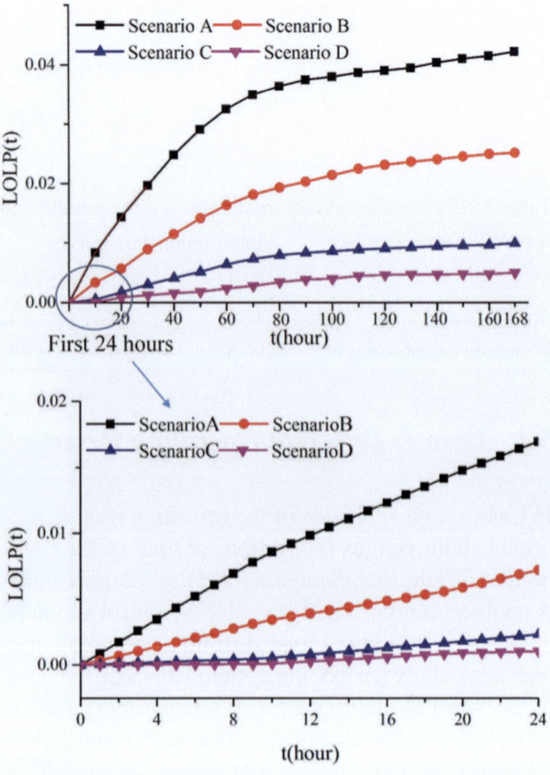

Fig. 11 System LOLP of case 1

Table 6 System costs for scenarios in case 1($)

Scenario	A	B	C	D
Generation	32,278,302	31,463,027	31,210,852	31,089,518
Reserve	2,630,146	1,871,678	1,235,416	1,189,572
Shifting	0	662,838	1,101,018	1,173,866
Curtailment	0	129,050	149,583	162,686
Interruption	5,440,792	1,444,604	316,508	169,484
Total cost	40,349,240	35,571,197	34,013,377	33,785,126

5.2 *Case 2: Uncertainties from VPPs*

In Case 2, it demonstrates three kinds of uncertainties, which are user engagement performance, ICS random failures and different load types.This case focuses on VPP uncertainties, which is divided into four categories for the level of participation, and the four categories mentioned in Table 7 are divided into Level 1 (full participation, 100%), Level 2 (intermediate state 1, 60%), Level 3 (intermediate state 2, 30%) and Level 4 (non-participation, 0%). The initial state is assumed at Level 1. For simplification of the simulation, the transition rates between different participation levels for each load state are assumed to be the same as shown in Table 7. The failure rate and repair rate of ICS are $8 * 10^{-4}$/h and 0.02/h [40].

The three scenarios of the operating reserve are shown in Table 8 below, where VPPs provide with various proportions of load shifting and load curtailment. Cases 1 and Case 2 only differ in the standard deviation on the proportion set.

The results of Case 2 in Table 9 clearly confirms that changes in uncertainties from the VPP cause changes in system reliability. The more load can be reduced, the lower EENS (a measure of system reliability) will get, which indicates a more reliable system, as well as lower total system cost.

Table 7 Transition rates between different participation levels of customers

Transition rates (/h)	Level 1	Level 2	Level 3	Level 4
Level 1	–	0.067	0.016	0.016
Level 2	0.359	–	0.185	0.038
Level 3	0.155	0.192	–	0.122
Level 4	0.018	0.008	0.013	–

Table 8 The mean values of proportions for shiftable load and curtailable load for different scenarios in case 2

Scenario	A	B	C
Load shifting	0.15	0.1	0.2
Load curtailment	0.15	0.2	0.1

Table 9 System EENS and costs for scenarios in case 2

Scenario	EENS (MWh)	Relative difference	Interruption cost ($)	Total cost ($)
A	440.12	0	1,734,953	35,857,945
B	392.59	−10.80%	1,520,108	35,646,100
C	467.72	6.27%	2,030,840	36,133,832

Table 10 System EENS and costs for scenarios in case 3

Scenario	EENS (MWh)	Relative difference	Shifting cost ($)	Total cost ($)
A	844.32	91.84%	549,524	37,671,374
B	484.60	10.10%	676,998	36,221,672
C	440.12	0	694,889	35,857,945

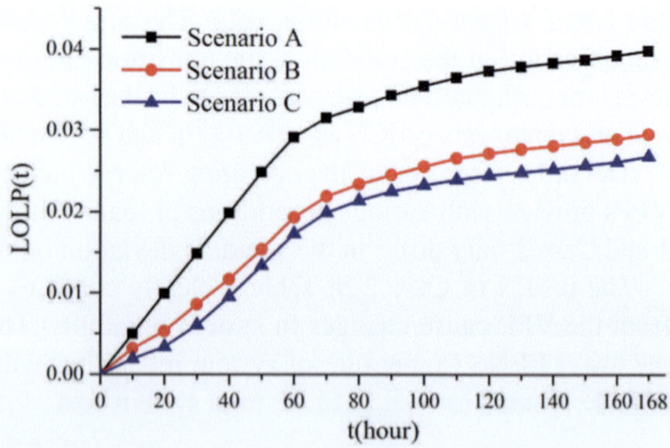

Fig. 12 System LOLP of case 3

5.3 *Case 3: Different Shiftable Time Periods*

In Case 3, fixed shifting capacity and fixed curtailment capacity are assumed at 15% and 15%, respectively. Three scenarios according to the different shiftable time is provided. In Scenario A, the load can be shifted only to the next one hour. In Scenario B, the load can be shifted to the next three hours. In Scenario C, the load can be shifted to the next five hours.

The system EENS for one week and system LOLP are presented in Table 10 and Fig. 12, respectively.

It can be observed from Table 10 and Fig. 12 that if the shiftable time periods are not limited, the system with longer shiftable time periods is more reliable. Moreover, the total system cost is lower while the shifting cost is higher. It is because that with longer shiftable time periods, the shiftable load is more likely to be transferred the other time periods when the system is in peak hours or contingency state. Total cost and shifting cost show opposite trends, which caused by different system states and thus different shifting target times.

5.4 *Case 4: Different Locations of VPP*

In Case 4, fixed shifting capacity and fixed curtailment capacity are assumed at 15% and 15%. Supposing a bus is "weak" and the two transmission lines connected to it

Table 11 System EENS and costs for scenarios in case 4

Scenario	EENS (MWh)	Relative difference	Interruption cost ($)	Total cost ($)
A	533.04	0	2,319,789	36,707,133
B	376.59	−29.35%	1,416,731	35,582,250

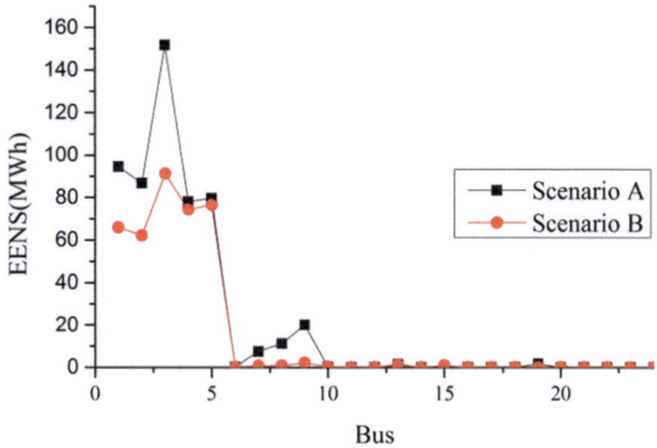

Fig. 13 EENS at different buses in case 4

are in "failure" state. Two scenarios are set up. In Scenario A, the VPPs distributed proportionally to the load at each bus. In Scenario B, the reserve capacity of VPPs is aggregated to weak bus.

The results of EENS which represent system reliability, and the system cost are shown in Table 11 and Fig. 13. The system reliability on the weak bus is lower than the normal state, which is related to the degree of line unclogging, making a direct impact on whether the reserve capacity can reach the bus. Meanwhile, it can be inferred that the system reliability of a VPP connected with a normal bus is related to the cost of interruption of the residential users. Lower interruption cost leads to higher system stability.

5.5 *Case 5: Load Model with Fixed Time Steps*

In Case 5, fixed shifting capacity and fixed curtailment capacity are assumed at 50% and 50% of the whole load, utilizing the load model with fixed time steps without uncertainties from VPPs.

The results of load model with fixed time steps present in Fig. 14. Comparing the results of the base scenario (0 curtailment and 0 shifting) in Case 1, the 5-h shifting load in Case 3, and Case 5, EENS for system reliability is lower. As the proportion of VPP participation increases, EENS continues to decrease, but stabilizes after a certain point. This case excludes the uncertainty of the duration of the load level. This means that the relationship between operating reserve capacity provided by VPPs and

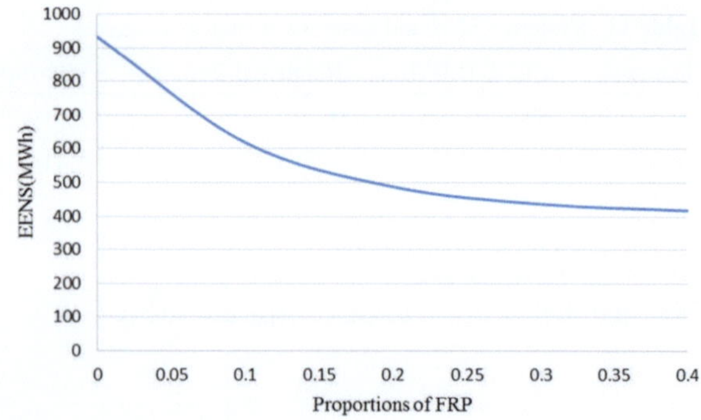

Fig. 14 System EENS with different proportions in case 5

system reliability will be uncertain when this factor is added. Power investment and development plans involving VPPs will likely yield different analytical emphases, and indicators of the effectiveness of operating reserve capacity provided by VPPs will need to be viewed on a case-by-case basis.

References

1. F.D. Galiana, F. Bouffard et al., Scheduling and pricing of coupled energy and primary, secondary, and tertiary reserves. Proc. IEEE **93**(11), 1970–1983 (2005)
2. "Glossary of terms," *Glossary of Terms Task Force, North American Electric Reliability Council* (1996) [Online] Available: ftp://www.nerc.com/pub/sys/all_updl/docs/pubs/glossv10.pdf
3. X. Li et al., A novel sensorless photovoltaic power reserve control with simple real-time MPP estimition. Electr. Power Syst. Reas. **34**(8), 7521–7531 (2019)
4. Y. Ding, W. Cui, et al., Multi-state operating reserve model of aggregate thermostatically-controlled-loads for power system short-term reliability evaluation. Appl. Energy **241**, 46–58 (2019)
5. X. Yan, H. Qiu, et al., Optimal configuration of a power grid system with a dynamic performance sharing mechanism. Reliab. Eng. Syst. Safety **193** (2020)
6. X. Qi et al., Short-term reliability assessment of generating systems considering DR reliability. IEEE Trans. Power Syst. **8**, 74371–74384 (2020)
7. B. Hu et al., Decision-dependent uncertainty modeling in power system operational reliability evaluations. IEEE Trans. Power Syst. **36**(6), 5708–5721 (2021)
8. G. Bao, Q. Li, J. Wang, et al., A comprehensive evaluation model of renewable energy benefits for power system planning. In *2019 IEEE 8th International Conference on Advanced Power System Automation and Protection (APAP)* (2019), pp. 986–990
9. Y. Yan, Y. Ding, et al., Operating reliability analysis of peaking generating units considering start-up failures and degradation. In *2016 Second International Symposium on Stochastic Models in Reliability Engineering, Life Science and Operations Management (SMRLO)* (Beer Sheva, 2016), pp. 168–171
10. Y. Ding, L. Cheng et al., Operational reliability evaluation of restructured power systems with wind power penetration utilizing reliability network equivalent and time-sequential simulation approaches. J. Mod. Power Syst. Clean Energy **2**(4), 329–340 (2014)
11. M. Sepehry, M. Heidari-Kapourchali, et al., A new algorithm for reliability assessment of distribution system based on reliability network equivalent technique. In *2018 IEEE International Conference on Probabilistic Methods Applied to Power Systems (PMAPS)* (2018), pp. 1–5

12. H. Jia, Y. Ding et al., Operating reliability evaluation of power systems considering flexible reserve provider in demand side. IEEE Trans. Smart Grid **10**(3), 3452–3464 (2019)
13. Y. Wang, V. Vittal et al., Composite system reliability evaluation with essential reliability services assessment of wind power integrated power systems. IEEE Open Access J. Power Energy **7**, 403–413 (2020)
14. A. Mehrtash, P. Wang et al., Reliability evaluation of power systems considering restructuring and renewable generators. IEEE Trans. Power Syst. **27**(1), 243–250 (2012)
15. A. Moshari, A. Ebrahimi et al., Short-term impacts of dr programs on reliability of wind integrated power systems considering demand-side uncertainties. IEEE Trans. Power Syst. **31**(3), 2481–2490 (2016)
16. Z. Zhou et al., Research about energy optimization management of large-scale charging piles for virtual power plant. In *2021 6th International Conference on Power and Renewable Energy (ICPRE)* (2021), pp. 1057–1062
17. S.A.A. Rizvi, A. Xin, et al., Optimal scheduling of virtual power plants utilizing wind power and electric vehicles. In *2021 11th International Conference on Power and Energy Systems (ICPES)* (2021), pp. 718–724
18. T. Da, C. Hou, et al, Equilibrium analysis on deep peak regulation market considering the participation of virtual power plants. In *2021 6th International Conference on Power and Renewable Energy (ICPRE)* (2021), pp. 1068–1072
19. T. Jin, M. Mechehoul, Ordering electricity via internet and its potentials for smart grid systems. IEEE Trans. Smart Grid **1**(3), 302–310 (2010)
20. Y. Xu, C. Singh, Power system reliability impact of energy storage integration with intelligent operation strategy. IEEE Trans. Smart Grid **5**(2), 1129–1137 (2014)
21. M. Carrión, Y. Dvorkin et al., Primary frequency response in capacity expansion with energy storage. IEEE Trans. Power Syst. **33**(2), 1824–1835 (2018)
22. L. Cheng, Y. Chang et al., Power system reliability assessment with electric vehicle integration using battery exchange mode. IEEE Trans. Sustain. Energy **4**(4), 1034–1042 (2013)
23. J. Choi, J. Lim et al., DSM considered probabilistic reliability evaluation and an information system for power systems including wind turbine generators. IEEE Trans. Smart Grid **4**(1), 425–432 (2013)
24. L. Ye, L. Zhao, S. Dong, et al., Dispatch in electric vehicles embedded virtual power plants considering safety constraints. In *2019 IEEE 3rd International Conference on Circuits, Systems and Devices (ICCSD)* (2019), pp. 143–147
25. A.M. Leite da Silva, L.A. da Fonscca Manso, et al., Pseudo-chronological simulation for composite reliability analysis with time varying loads. IEEE Trans. Power Syst. **15**(1), 73–80 (2000)
26. R. Billinton, C. Singh, System load representation in generating capacity reliability studies part I model formulation and analysis. IEEE Trans. Power Appar. Syst. **PAS-91**(5), 2125–2132
27. Y. Wang, Q. Chen, C. Kang et al., Clustering of electricity consumption behavior dynamics toward big data applications. IEEE Trans. Smart Grid **7**(5), 2437–2447 (2016)
28. F.F.C. Veliz, C.L.T. Borges et al., A comparison of load models for composite reliability evaluation by nonsequential Monte Carlo simulation. IEEE Trans. Power Syst. **25**(2), 649–656 (2010)
29. K.R. Davis, C.M. Davis et al., A cyber-physical modeling and assessment framework for power grid infrastructures. IEEE Trans. Smart Grid **6**(5), 2464–2475 (2015)
30. W. Kuo, M. Zuo, *Optimal Reliability Modeling Principles and Applications* (John Wiley & Sons, 2003)
31. R. Billinton, N.R. *Allan, Reliability Evaluation of Engineering Systems, Concepts and Techniques*, 2nd ed. (Plenum Press, pp. 377–378, 1992)
32. Y. Ding, P. Wang, Reliability and price risk assessment of a restructured power system with hybrid market structure. IEEE Trans. Power Syst. **21**(1), 108–116 (2006)
33. A. Lisnianski, D. Elmakias et al., A multi-state Markov model for a short-term reliability analysis of a power generating unit. Reliab. Eng. Syst. Saf. **98**(1), 1–6 (2012)

34. Y. Bapin, M. Mussard et al., Estimation of operating reserve capacity in interconnected systems with variable renewable energy using probabilistic approach. IEEE Int. Conf. Probabil. Meth. Appl. Power Syst. (PMAPS) **2018**, 1–6 (2018)
35. R. Karki, Po Hu, et al., A simplified wind power generation model for reliability evaluation. IEEE Trans. Energy Convers. **21**(2), 533–540 (2006)
36. P. Wang, Y. Ding et al., Reliability assessment of restructured power systems using optimal load shedding technique. IET Gener. Transm. Distrib. **3**(7), 628–640 (2009)
37. R. Billinton, S. Kumar, N. Chowdhhury et al., A reliability test system for educational purposes-basic data. IEEE Trans. Power Syst. **4**(3), 1238–1244 (1989)
38. G. Tollefson, R. Billinton et al., A Canadian customer survey to assess power system reliability worth. IEEE Trans. Power Syst. **9**(1), 443–450 (1994)
39. M.J. Sullivan, M.G. Mercurio, et al., *Estimated Value of Service Reliability for Electric Utility Customers in the United States* (LBNL Research Project Final Report, 2009)
40. M. Pantel, D.S. Kirschen, Assessing the effect of failures in the information and communication infrastructure on power system reliability. In *2011 IEEE/PES Power Systems Conference and Exposition* (Phoenix, AZ, 2011), pp. 1-7

Chapter 9
Reliable Operation of Power Systems Integrated with Virtual Power Plants and Wind Power Considering Cyber Malfunctions

Building on the previous chapters, this chapter will further enrich the reliable operation model to analyze the impact of VPPs on power systems incorporating wind power. The reliable operation model is for VPPs with typical hierarchical decentralized control infrastructures, taking cyber network and communication failures into account. Then, the capacity of the operating reserves provided by VPPs is presented considering the stochastic behavior characteristics of DERs. Moreover, a wind power generation system is added to the conventional generation system, and the reliability assessment by a multi-state model is mainly based on some reliability indicators of the optimal power flow. Finally, a modified IEEE RTS is used to verify the above model that the operational reserve of VPPs can obviously improve the reliability of power systems, but it is affected by some conditions of cyber network or communication attacks, and several factors of VPPs.

1 Introduction

1.1 Background

The Internet of Things (IoT) technology is booming and has been applied to make the interaction between cyber network and physical systems more complex. With the application of IoT within the power sector, smart grids allow special network structures and physical structures within power systems to interoperate, through emerging normative technologies and protocols [1, 2]. The diffusion and application of emerging information technologies allows VPPs to aggregate demand-side resources, namely distributed energy resources (DERs), as operational reserves to support the normal operation of power systems [3]. By participating in the power markets through bilateral contracts or demand response, VPPs can have controllability on DERs. Taking these two classical scenarios for example, interruptible loads

H. Jia et al., *Business Models and Reliable Operation of Virtual Power Plants*,
https://doi.org/10.1007/978-981-19-7846-3_9

can provide capacity through bilateral contracts, or they can participate in competitive bidding in a demand response manner [4–6].

Emphasis on network and information technology development should be accompanied by possible risky failures, such as cyber attacks and communication delays, which can seriously affect the stable operation of VPPs and thus bring down power systems [7]. In 2015, the Ukrainian power grid had a blackout caused by cyber attacks [8]. The operational instability of power systems also comes from the fluctuating and intermittent characteristics of wind power systems and the stochastic characteristics of VPPs [9]. Therefore, a quantitative study of reliability assessment of restructured power systems consisting of wind power generation and VPPs is necessary [10].

1.2 Related Works

There has been research which has considered the effect of cyber malfunctions in power system reliability assessment, for example, literature [11] focused on the evaluation of network links for active cyber physical distribution systems. A modified Bayesian attack graph model was utilized for studying wind farm energy management systems subjected to a cyber attack on a power system [12]. All the above literature used simulation techniques, and there was also literature that did not consider cyber attacks but used the sequential simulation method for power system reliability evaluation [13, 14]. Although the simulation technique is applicable for reliability assessment under general power system operating conditions, the results still have limitations with accuracy [15]. In summary, this chapter proposes an analytic reliability assessment method considering cyber malfunctions, which is integrated with VPPs and wind power.

There are stochastic processes in wind power systems and corresponding stochastic processes in VPPs aggregation, and a multi-state model can describe performances of such systems well [16, 17]. However, the number and complexity of system states increase significantly with the increase of system components. Even though the constituent structures of power systems are relatively simple, the system states are very diverse and need to be studied in depth for modeling stochastic processes.

Researchers have used different quantitative methods for relatively accurate reliability analysis, but they have their limitations. Markov process analysis methods have been widely utilized [18], but traditional analytical models can occur the curse of dimensionality due to the messy of system states. The universal generating function (UGF) [19] method can describe the performance distribution of a multi-state system and reduce the computational complexity [20–23]. This method was further extended to become the *Lz* transform method to evaluate the system dynamic reliability considering time-varying factors [17, 20]. However, VPPs' failures due to cyber attacks and communication delays are not included in the evaluation.

There are also studies related to cyber network and communication problems occurring in power systems specifically. In terms of cyber malfunctions occurring in the power grid, for example, there are studies focusing on line-outage positions detection in transmission lines [24], cascading failure process between the grid and the cyber network through stochastic failure model [25], and classification of cyber network attack strategies to study attack rates and defense [26]. In communication, for example, research on the system control impact from communication delays occurring in islanded microgrids [27], communication delays act on distributed frequency regulation [28], and cyber-physical model is proposed for physical component interference in power grid infrastructures [29]. And the above-mentioned studies related to cyber networks and communication malfunctions did not address the reliability assessment of wind power systems.

In [30], the *Lz* transform approach, which can simplify complex algorithms, was used to implement the process of power system reliability assessment by building a multi-state model that includes factors of cyber attacks and communication failures. The multi-state model was first developed for the wind power system and reserve capacity from demand side, where DERs are considered with a hierarchical decentralized control infrastructure. The reliability indicators on the basis of load curtailment were used to assess reliability by performing optimal power flow for different system states. In [30], an innovative multi-state assessment model was used to evaluate the reliability of power systems, and the *Lz* transform technique was used to simplify system complexity and arithmetic processes, enabling dynamic system evaluation. Reserve capacity modeling from demand side considered stochastic characteristics along with ICT infrastructure, and expanded the analysis of a typical hierarchical decentralized control infrastructure in the demand-side reliability modeling process in [30], to include cyber network attacks and communication delays as influencing factors. Moreover, considering cyber network failures, on the basis of optimal power flow, a time-varying characteristic reliability metric is selected to analyze the entire power system containing DERs and wind power systems. Extended from [30], in this chapter, DERs are presented as VPPs. The overview of the operating reliability evaluation framework for power systems with VPPs considering cyber malfunction is illustrated in Fig. 1.

2 Multi-state Reserve Capacity Model for VPPs Considering Cyber Malfunctions

2.1 Reliability Model for the Cyber System with a Hierarchical Decentralized Control Framework

The regulation of aggregation of DERs within VPPs is achieved through a cyber infrastructure with a hierarchical decentralized control framework. The control framework shown in Fig. 2 consists of a control center and several local controllers (multiple Q_i) that are embedded in a number of load sectors.

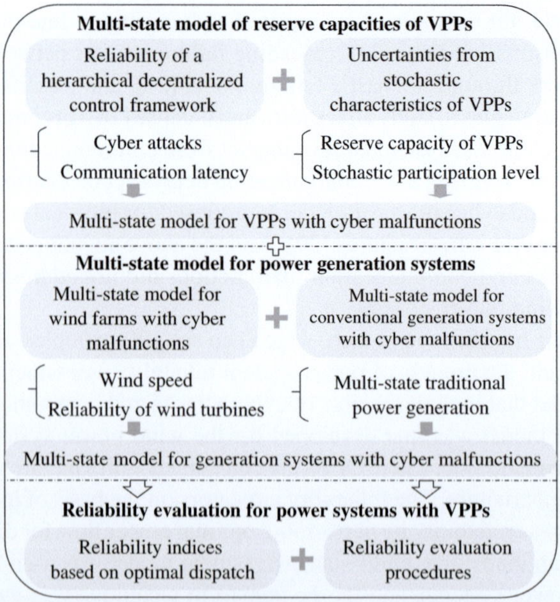

Fig. 1 Overview of the operating reliability evaluation framework for power systems with VPPs considering cyber malfunctions

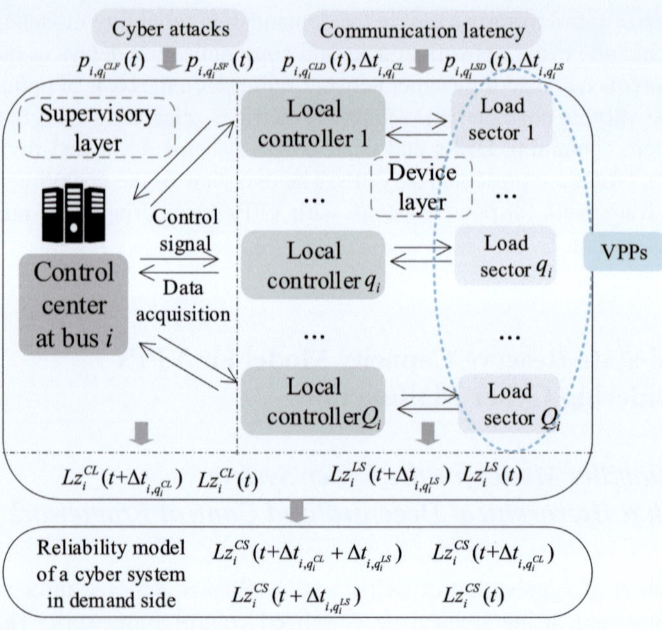

Fig. 2 A hierarchical decentralized control framework of the cyber system for bus i

The control strategy in the demand-side cyber network system contains two decision layers, the supervisory layer which is responsible for sending control signals, and the device layer which receives the signals, as shown in Fig. 2. In the control process, after the control center located in the supervisory layer sending signals to the controllers, in turn, the VPPs in the device layer receive these signals and make response according to the signals from controllers. Inevitably, the attacks to cyber networks and communication delay failures occur during signal transmission. If the supervisory layer is invaded by cyber attacks, the whole system will be down and thus unable to regulate the VPPs. The VPPs may also respond slowly due to the communication delay situations, miss the response time and generate losses. Similar to the above, cyber malfunctions are important unexpected factors that influence system reliability.

The reliability assessment model containing the cyber system is built in two parts, on the one hand, from the control center in the supervisory layer to the local controller q_i, and on the other hand, the local controller q_i in the device layer to the load sector, where VPPs are located in the load sector q_i on the bus i.

The next step will describe the situation whether the cyber system failures take place as binary-state models, that is to say, systems without cyber attacks is in the "1" state, and systems suffering cyber attacks and failing to function is in the "0" state. Therefore, for the time t, the Lz transforms for cyber attacks of the supervisory layer from the control center to the local controller q_i and the device layer from the local controller q_i to the corresponding load sector can be formulated in (1)–(2), respectively. z is a complex variable providing a comprehensive method for the system state enumeration and substituting complicated combinational algorithm.

$$Lz_{i,q}^{CLF}(t) = p_{i,q_i^{CLF}}(t) \cdot z^0 + \left(1 - p_{i,q_i^{CLF}}(t)\right) \cdot z^1 \tag{1}$$

$$Lz_{i,q}^{LSF}(t) = p_{i,q_i^{LSF}}(t) \cdot z^0 + \left(1 - p_{i,q_i^{LSF}}(t)\right) \cdot z^1 \tag{2}$$

Moreover, for the time t, taking the supervisory layer for example, if communication latency occurs for $\Delta t_{i,q_i^{CL}}$ with a probability $p_{i,q_i^{CLD}}(t)$, the functioning of the cyber system in the time $(t + \Delta t_{i,q_i^{CL}})$ is affected; If the communication condition is completely normal, the cyber system will also operate normally and is not affected by the delay. The Lz transforms for both layers are shown in Eqs. (3)–(4).

$$\begin{cases} Lz_{i,q}^{CLD}(t + \Delta t_{i,q_i^{CL}}) = p_{i,q_i^{CLD}}(t) \cdot z^1 \\ Lz_{i,q}^{CLD}(t) = \left(1 - p_{i,q_i^{CLD}}(t)\right) \cdot z^1 \end{cases} \tag{3}$$

$$\begin{cases} Lz_{i,q}^{LSD}(t + \Delta t_{i,q_i^{LS}}) = p_{i,q_i^{LSD}}(t) \cdot z^1 \\ Lz_{i,q}^{LSD}(t) = \left(1 - p_{i,q_i^{LSD}}(t)\right) \cdot z^1 \end{cases} \tag{4}$$

Considering both cyber attacks and communication latency, the Lz transforms for two layers are formulated in (5)–(6), respectively.

$$\begin{cases} Lz_{i,q}^{CL}(t+\Delta t_{i,q_i^{CL}}) = p_{i,q_i^{CLD}}(t) \cdot Lz_{i,q}^{CLF}(t+\Delta t_{i,q_i^{CL}}) \\ Lz_{i,q}^{CL}(t) = \left(1 - p_{i,q_i^{CLD}}(t)\right) \cdot Lz_{i,q}^{CLF}(t) \end{cases} \tag{5}$$

$$\begin{cases} Lz_{i,q}^{LS}(t+\Delta t_{i,q_i^{LS}}) = p_{i,q_i^{LSD}}(t) \cdot Lz_{i,q}^{LSF}(t+\Delta t_{i,q_i^{LS}}) \\ Lz_{i,q}^{LS}(t) = \left(1 - p_{i,q_i^{LSD}}(t)\right) \cdot Lz_{i,q}^{LSF}(t) \end{cases} \tag{6}$$

For the load sector q_i at bus i, the reliability model of its cyber systems for both layers has four forms with communication latencies in both layers, in the supervisory layer, in the device layer, and no latency. The corresponding Lz transforms can be formulated using series operator Ω_s in (7)–(10), respectively.

The Lz transform for the reliability model for cyber systems with communication latencies in both layers is presented in (7).

$$\begin{aligned} & Lz_{i,q}^{CS}(t + \Delta t_{i,q_i^{CL}} + \Delta t_{i,q_i^{LS}}) \\ & = \Omega_s\left(Lz_{i,q}^{CL}(t + \Delta t_{i,q_i^{CL}}), Lz_{i,q}^{LS}(t + \Delta t_{i,q_i^{LS}})\right) \\ & = \Omega_s\begin{pmatrix} p_{i,q_i^{CLD}}(t) \cdot \left(p_{i,q_i^{CLF}}(t + \Delta t_{i,q_i^{CL}}) \cdot z^0 + \left(1 - p_{i,q_i^{CLF}}(t + \Delta t_{i,q_i^{CL}})\right) \cdot z^1\right), \\ p_{i,q_i^{LSD}}(t) \cdot \left(p_{i,q_i^{LSF}}(t + \Delta t_{i,q_i^{LS}}) \cdot z^0 + \left(1 - p_{i,q_i^{LSF}}(t + \Delta t_{i,q_i^{LS}})\right) \cdot z^1\right) \end{pmatrix} \end{aligned} \tag{7}$$

The Lz transform for the reliability model for cyber systems with communication latency in only supervisory layer is presented in (8).

$$\begin{aligned} & Lz_{i,q}^{CS}(t + \Delta t_{i,q_i^{CL}}) \\ & = \Omega_s\left(Lz_{i,q}^{CL}(t + \Delta t_{i,q_i^{CL}}), Lz_{i,q}^{LS}(t)\right) \\ & = \Omega_s\begin{pmatrix} p_{i,q_i^{CLD}}(t) \cdot \left(p_{i,q_i^{CLF}}(t + \Delta t_{i,q_i^{CL}}) \cdot z^0 + \left(1 - p_{i,q_i^{CLF}}(t + \Delta t_{i,q_i^{CL}})\right) \cdot z^1\right), \\ \left(1 - p_{i,q_i^{LSD}}(t)\right) \cdot \left(p_{i,q_i^{LSF}}(t) \cdot z^0 + \left(1 - p_{i,q_i^{LSF}}(t)\right) \cdot z^1\right) \end{pmatrix} \end{aligned} \tag{8}$$

The Lz transform for the reliability model for cyber systems with communication latency in only device layer is presented in (9).

$$\begin{aligned} & Lz_{i,q}^{CS}(t+\Delta t_{i,q_i^{LS}}) \\ & = \Omega_s\Big(Lz_{i,q}^{CL}(t), Lz_{i,q}^{LS}(t+\Delta t_{i,q_i^{LS}})\Big) \\ & = \Omega_s\left(\begin{array}{l} \Big(1-p_{i,q_i^{CLD}}(t)\Big)\cdot\Big(p_{i,q_i^{CLF}}(t)\cdot z^0+\Big(1-p_{i,q_i^{CLF}}(t)\Big)\cdot z^1\Big), \\ p_{i,q_i^{LSD}}(t)\cdot\Big(p_{i,q_i^{LSF}}(t+\Delta t_{i,q_i^{LS}})\cdot z^0+\Big(1-p_{i,q_i^{LSF}}(t+\Delta t_{i,q_i^{LS}})\Big)\cdot z^1\Big) \end{array}\right) \end{aligned} \tag{9}$$

Specifically, if no communication latency occurs which indicates $\Delta t_{i,q_i^{CL}} = \Delta t_{i,q_i^{LS}} = 0$, the corresponding Lz transform for a cyber system in this scenario is formulated in (10).

$$\begin{aligned} Lz_{i,q}^{CS}(t) & = \Omega_s\Big(Lz_{i,q}^{CL}(t), Lz_{i,q}^{LS}(t)\Big) \\ & = \Omega_s\left(\begin{array}{l} \Big(1-p_{i,q_i^{CLD}}(t)\Big)\cdot\Big(p_{i,q_i^{CLF}}(t)\cdot z^0+\Big(1-p_{i,q_i^{CLF}}(t)\Big)\cdot z^1\Big), \\ \Big(1-p_{i,q_i^{LSD}}(t)\Big)\cdot\Big(p_{i,q_i^{LSF}}(t)\cdot z^0+\Big(1-p_{i,q_i^{LSF}}(t)\Big)\cdot z^1\Big) \end{array}\right) \end{aligned} \tag{10}$$

2.2 *Uncertainties from Stochastic Characteristics of VPPs*

Setting up power systems with VPPs is more flexible and complex because VPPs have stochastic characteristics, such as the thermodynamic characteristics of interruptible loads in VPPs and other physical characteristics. Therefore, a multi-state model is established to describe the operating reserve capacity provided by VPPs.

As presented in Fig. 3, the multi-state model of available reserve capacity provided by VPPs considering stochastic characteristics form the physical characteristics of VPPs and chaptericipation levels of customers.

Let $OR_{i,j_i^{VPP}}, j_i^{VPP} = 1, \ldots, K_i^{VPP}$ be the reserve capacity from a VPP in the state j_i^{VPP} at bus i. The number of discrete states K_i^{VPP} is determined by the maximal reserve capacity and the accuracy of multi-state model. An appropriate value of K_i^{VPP} is supposed to be determined for balancing the accuracy and the complexity. Let $p_{i,j_i^{VPP}}, j_i^{VPP} = 1, \ldots, K_i^{VPP}$ be the probability of the reserve capacity of a VPP in the state j^{VPP} at bus i. Therefore, the Lz representation of the reserve capacity of VPPs is formulated in (11).

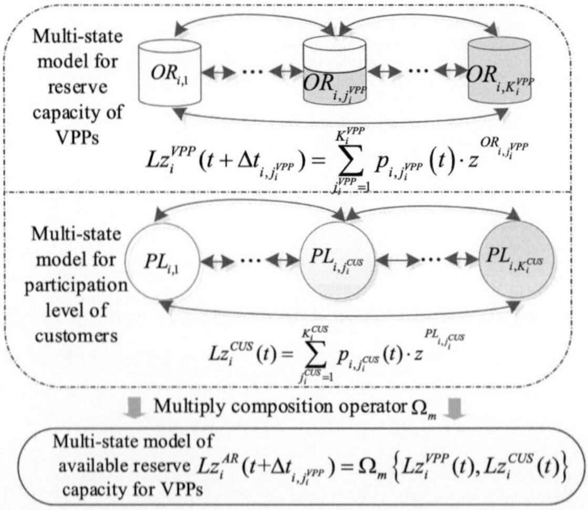

Fig. 3 A multi-state model for available reserve capacity of VPPs

$$Lz_i^{VPP}(t + \Delta t_{i,j_i^{VPP}}) = \sum_{j_i^{DSR}=1}^{K_i^{VPP}} p_{i,j_i^{VPP}}(t) \cdot z^{OR_{i,j_i^{VPP}}} \tag{11}$$

The reserve capacity of VPPs is not only closely related to the physical characteristics mentioned, but also related to the consumption characteristics and incentive degree of DERs on the demand side. For example, during peak hours in summer, interruptible loads are not necessarily willing to provide reserve capacity, which directly reduces the reserve capacity of VPPs. Another example, when power users are not attracted to the demand response price, especially those with low power costs, they will refuse to participate in aggregation. The corresponding multi-state model is set again for this case where the level states and transition rates are determined from the historical data.

In Fig. 3, $PL_{i,j_i^{CUS}}, j_i^{CUS} = 1, \ldots, K_i^{CUS}$ is the participation level of customers for the state j_i^{CUS} at bus i. According to these transition rates $\lambda_{j_i^{CUS},k_i^{CUS}}$ and the initial state of the participation level, $p_{i,j_i^{CUS}}(t), j_i^{CUS} = 1, \ldots, K_i^{CUS}$ can be solved for obtaining the probability for participation level of the state j_i^{CUS} at bus i by solving differential equations. The Lz transform to represent the participation level distribution can be defined as the following polynomial.

$$Lz_i^{CUS}(t) = \sum_{j_i^{CUS}=1}^{K_i^{CUS}} p_{i,j_i^{CUS}}(t) \cdot z^{PL_{i,j_i^{CUS}}} \tag{12}$$

Considering both original reserve capacity from VPPs and participation level of customers, the Lz transform of the available reserve capacity from VPPs can be obtained utilizing the multiply composition operator Ω_m as following.

$$\begin{aligned}
Lz_i^{AR}(t+\Delta t_{i,j_i^{VPP}}) &= \Omega_m\left\{Lz_i^{VPP}(t), Lz_i^{CUS}(t)\right\} \\
&= \Omega_m\left\{\sum_{j_i^{VPP}=1}^{K_i^{VPP}} p_{i,j_i^{VPP}}(t)\cdot z^{OR_{i,j_i^{VPP}}}, \sum_{j_i^{CUS}=1}^{K_i^{CUS}} p_{i,j_i^{CUS}}(t)\cdot z^{PL_{i,j_i^{CUS}}}\right\} \\
&= \sum_{j_i^{VPP}=1}^{K_i^{VPP}}\sum_{j_i^{CUS}=1}^{K_i^{CUS}} p_{i,j_i^{VPP}}(t)\cdot p_{i,j_i^{CUS}}(t)\cdot z^{OR_{i,j_i^{VPP}}\cdot PL_{i,j_i^{CUS}}} \\
&= \sum_{j_i^{AR}=1}^{K_i^{AR}} p_{i,j_i^{AR}}(t)\cdot z^{ARC_{i,j_i^{AR}}}
\end{aligned} \tag{13}$$

2.3 *Multi-state Model for VPPs*

Considering the previous factors, the reserve capacity of VPPs discussed in this chapter depends on the cyber network status, the total load capacity of the interruptible load and the participation level of customers.

Combined with the binary-state model of cyber attacks and considering the communication latency factor, the actual reserve capacity of VPP is represented by (14).

$$\begin{aligned}
&Lz_i^{AAR}(t+\Delta t_{i,q_i^{CL}}+\Delta t_{i,q_i^{LS}}+\Delta t_{i,j_i^{VPP}}) \\
&= \Omega_m\left\{\sum_{q=1}^{Q} Lz_{i,q}^{CS}(t+\Delta t_{i,q_i^{CL}}+\Delta t_{i,q_i^{LS}}), Lz_i^{AR}(t+\Delta t_{i,j_i^{VPP}})\right\} \\
&= \sum_{j_i^{AAR}=1}^{K_i^{AAR}} p_{i,j_i^{AAR}}(t)\cdot z^{AARC_{i,j_i^{AAR}}}
\end{aligned} \tag{14}$$

3 Multi-state Model for Power Generation Systems with Cyber Malfunctions

Considering the interconnection between cyber systems and physical systems in power generation, the influence of cyber malfunctions due to cyber attacks and communication latency on wind power and conventional generating units are modeled in this section. Figure 4 presents the reliability model for power generation systems with cyber malfunctions.

3.1 Multi-state Model for Wind Farms with Cyber Malfunctions

This section establishes a multi-state model considering network faults. The output power of wind farms is determined not only by wind turbines, but also by cyber network systems.

(1) Multi-state model for a wind turbine

In terms of wind turbine related factors, the uncertain wind speeds and corresponding realistic physical conditions jointly determine the output power of the wind farm.

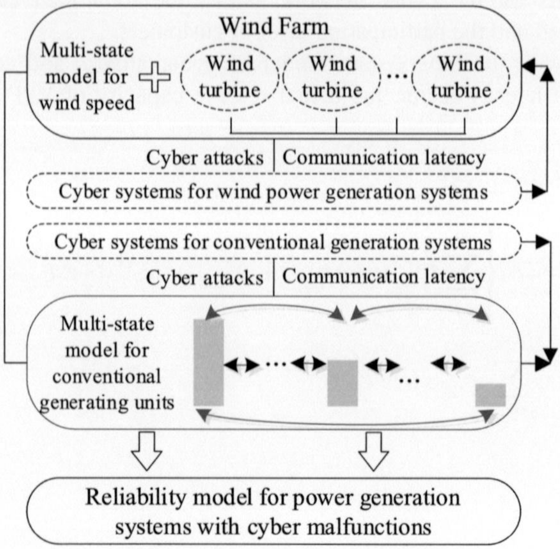

Fig. 4 Reliability model for power generation systems with cyber malfunctions

It is common practice to establish Markov process model for the time-varying properties of stochastic wind speeds. During the normal working state period of power systems, the two variables related to wind speeds determine the state probability of the future wind speed, which are the current and future wind speed data, respectively. In Markov process, the multi-state wind speed and state transition rates can be collected by operational data analysis.

Let $S_{i,j_i^{WS}}, j_i^{WS} = 1, \ldots, K_i^{WS}$ be the wind speed for bus i at time t which is a stochastic variable. The number of discrete states K_i^{WS} is determined by the accuracy of multi-state model. In the Markov process model for wind speeds, differential equations are utilized for solving the state probability $p_{i,j_i^{WS}}$ as formulated in (15), under the initial conditions in (16).

$$\frac{dp_{i,j_i^{WS}}(t)}{dt} = \left[\sum_{k_i^{WS}=1, k_i^{WS} \neq j_i^{WS}}^{k} p_{i,k_i^{WS}}(t) \cdot \lambda_{k_i^{WS}, j_i^{WS}} \right] - p_{i,j_i^{WS}}(t) \sum_{j_i^{WS}=1, k_i^{WS} \neq j_i^{WS}}^{k} \lambda_{j_i^{WS}, k_i^{WS}} \tag{15}$$

$$p_{i,j_i^{WS}}(t_0) = 1, p_{i,k_i^{WS}}(t_0) = 0 \quad j_i^{WS} \neq k_i^{WS} \tag{16}$$

According to the corresponding relationship [9] between wind power and wind speed, the power outputs under different wind speeds can be represented as $SP_{i,j_i^{WSP}}, j_i^{WSP} = 1, \cdots, K_i^{WSP}$ with a probability $p_{i,j_i^{WSP}}$. Hence, considering the stochastic characteristics of wind speed, the Lz transform for wind speed at bus i can be formulated in (17).

$$Lz_i^{WSP}(t) = \sum_{j_i^{WSP}=1}^{K_i^{WSP}} p_{i,j_i^{WSP}}(t) \cdot z^{SP_{i,j_i^{WSP}}} \tag{17}$$

The normal operation of wind turbine also requires reliable support of physical parts. The well-functioning state of wind turbine is rated power. Random failures of wind turbines are described in a binary model as (18), with normal operation state "1" and failure state "0".

$$Lz_{i,u}^{WT}(t) = p_{i,u_i^{WT}}(t) \cdot z^1 + \left(1 - p_{i,u_i^{WT}}(t)\right) \cdot z^0 \tag{18}$$

Therefore, considering both of wind speeds and reliability of a wind turbine, the Lz representation for the output of wind turbine u is formulated in (19) utilizing the multiply operator Ω_m.

$$\begin{aligned} Lz_{i,u}^{WO}(t) &= \Omega_m\left\{Lz_i^{WSP}(t), Lz_{i,u}^{WT}(t)\right\} \\ &= \Omega_m\left\{\sum_{j_i^{WSP}=1}^{K_i^{WSP}} p_{i,j_i^{WSP}}(t)\cdot z^{SP_{i,j_i^{WSP}}}, p_{i,u_i^{WT}}(t)\cdot z^1 + \left(1-p_{i,u_i^{WT}}(t)\right)\cdot z^0\right\} \\ &= \sum_{j_i^{WSP}=1}^{K_i^{WSP}} p_{i,j_i^{WSP}}(t)\cdot p_{i,u_i^{WT}}(t)\cdot z^{SP_{i,j_i^{WSP}}} \\ &\quad + \sum_{j_i^{WSP}=1}^{K_i^{WSP}} p_{i,j_i^{WSP}}(t)\cdot\left(1-p_{i,u_i^{WT}}(t)\right)\cdot z^0 \\ &= \sum_{j_i^{WO}=1}^{K_i^{WO}} p_{i,j_i^{WO}}^{u}(t)\cdot z^{WPO^u_{i,j_i^{WO}}} \end{aligned} \tag{19}$$

(2) Multi-state model for a wind farm considering cyber systems

For a wind farm at bus i consisting of various wind turbines, the Lz transform of power output of the wind farm can be presented utilizing parallel operator Ω_p in (20).

$$\begin{aligned} Lz_i^{WF}(t) &= \Omega_p\left\{Lz_{i,1}^{WO}(t),\ldots,Lz_{i,u}^{WO}(t),\ldots\right\} \\ &= \Omega_p\left\{\ldots,\sum_{j_i^{WO}=1}^{K_i^{WO}} p_{i,j_i^{WO}}^{u}(t)\cdot z^{WPO^u_{i,j_i^{WO}}},\ldots\right\} = \sum_{j_i^{WF}=1}^{K_i^{WF}} p_{i,j_i^{WF}}(t)\cdot z^{WFP_{i,j_i^{WF}}} \end{aligned} \tag{20}$$

When the cyber malfunctions factors of the wind turbine are included, the Lz transform based on the binary model of the wind turbine for bus i is presented in (21).

$$Lz_i^{WC}(t+\Delta t_i^{WC}) = p_i^{WC}(t)\cdot z^1 + \left(1-p_i^{WC}(t)\right)\cdot z^0 \tag{21}$$

Therefore, the available power generation of a wind farm at bus i considering random malfunctions of cyber systems can be represented utilizing the multiply operator Ω_m in the following polynomial.

$$\begin{aligned} Lz_i^{AW}(t+\Delta t_i^{WC}) &= \Omega_m\left\{Lz^{wf}(t), Lz_i^{WC}(t+\Delta t_i^{WC})\right\} \\ &= \sum_{j_i^{WF}=1}^{K_i^{WF}} p_{i,j_i^{WF}}(t)\cdot p_i^{WC}(t)\cdot z^{WFP_{i,j_i^{WF}}} \\ &+ \sum_{j_i^{WF}=1}^{K_i^{WF}} p_{i,j_i^{WF}}(t)\cdot\left(1-p_i^{WC}(t)\right)\cdot z^0 \\ &= \sum_{j_i^{AW}=1}^{K_i^{AW}} p_{i,j_i^{AW}}(t)\cdot z^{AWP_{i,j_i^{AW}}} \end{aligned} \tag{22}$$

3.2 *Multi-state Model for Conventional Generation Systems with Cyber Malfunctions*

The random failures and component loss of conventional generator units will reduce power generation. Conventional power generation systems also have the stochastic characteristics, and it is difficult to capture the continuous state procession. Using Lz transform to establish the multi-state model of traditional generation systems is formulated as (23), the basic data can be obtained by analyzing the historical operation data of the units.

$$Lz_{i,v}^{CG}(t) = \sum_{j_{i,v}^{CG}=1}^{K_{i,v}^{CG}} p_{i,j_{i,v}^{CG}}(t)\cdot z^{CGP_{i,j_{i,v}^{CG}}} \tag{23}$$

For the conventional generation system at bus i, the multi-state model for the generation system based on Lz transform using parallel operator Ω_p is presented in (24).

$$Lz_i^{GS}(t) = \Omega_p\left\{Lz_{i,1}^{CG}(t), \ldots, Lz_{i,v}^{CG}(t), \ldots\right\}$$
$$= \Omega_p\left\{\ldots, \sum_{j_{i,v}^{CG}=1}^{K_{i,v}^{CG}} p_{i,j_{i,v}^{CG}}(t) \cdot z^{CGP_{i,j_{i,v}^{CG}}}, \ldots\right\}$$
$$= \sum_{j_i^{GS}=1}^{K_i^{GS}} p_{i,j_i^{GS}}(t) \cdot z^{GSP_{i,j_i^{GS}}} \tag{24}$$

When the cyber malfunctions factors of the traditional generation systems are included, the Lz transform based on the binary-state model and the reliability model of the traditional generation systems are presented in (25)–(26).

$$Lz_i^{GC}(t+\Delta t_i^{GC}) = p_i^{GC}(t) \cdot z^1 + \left(1 - p_i^{GC}(t)\right) \cdot z^0 \tag{25}$$

$$Lz_i^{AC}(t+\Delta t_i^{GC}) = \Omega_m\left\{Lz_i^{GS}(t), Lz_i^{GC}(t+\Delta t_i^{GC})\right\}$$
$$= \sum_{j_i^{GS}=1}^{K_i^{GS}} p_{i,j_i^{GS}}(t) \cdot p_i^{GC}(t) \cdot z^{GSP_{i,j_i^{GS}}}$$
$$+ \sum_{j_i^{GS}=1}^{K_i^{GS}} p_{i,j_i^{GS}}(t) \cdot \left(1 - p_i^{GC}(t)\right) \cdot z^{GSP_{i,j_i^{GS}}}$$
$$= \sum_{j_i^{AC}=1}^{K_i^{AC}} p_{i,j_i^{AC}}(t) \cdot z^{ACP_{i,j_i^{AC}}} \tag{26}$$

3.3 *Multi-state Model for Power Generation Systems with Cyber Malfunctions*

Multi-state models for wind power generation systems and conventional power generation systems have been established, and using parallel operator Ω_p can represent multi-state models of cyber attacks and communication delays as (27).

$$Lz_i^{G}(t + \Delta t_i^{WC} + \Delta t_i^{GC}) = \Omega_p\left\{Lz_i^{AW}(t+\Delta t_i^{WC}), Lz_i^{AC}(t+\Delta t_i^{GC})\right\}$$
$$= \sum_{j_i^{G}=1}^{K_i^{G}} p_{i,j_i^{G}}(t) \cdot z^{PG_{i,j_i^{G}}} \tag{27}$$

4 Reliability Evaluation of Power Systems with VPPs Considering Cyber Malfunctions

4.1 Reliability Indices

The Lz transform considering the cyber malfunctions has been established and obtained the reserve capacity of VPPs, then the regulation of an optimal power flow operator Ω_{opf} can realize the load curtailment of the whole system. For the load curtailment, Eq. (28) shows the Lz transform at but i with K states of a N-bus power system.

$$\begin{aligned} Lz_i^{LC}(t) &= \Omega_{opf}\left\{Lz_i^G(t), Lz_i^{AAR}(t)\right\} \\ &= \sum_{i=1}^{N}\sum_{j_i^G=1}^{K_i^G}\sum_{j^{TL}=1}^{K^{TL}}\sum_{j_i^{AAR}=1}^{K_i^{AAR}} p_{i,j_i^G}(t)\cdot p_{j^{TL}}(t)\cdot p_{i,j_i^{AAR}}(t)\cdot z^{opf\{GC_{i,j_i^G}, AARC_{i,j_i^{AAR}}\}} \\ &= \sum_{j=1}^{K} p_{j_i}(t)\cdot z^{LC_{j_i}(t)} \end{aligned} \tag{28}$$

The optimal power flow composition operator in (28) is defined as an optimization model to determine load curtailment for system state j at bus i, which is formulated in (29)–(33).

$$\min \quad f_j = \sum_{i=1}^{N} LC_{j_i}(t) \tag{29}$$

The objection function in (29) is to minimize the total load curtailment for the system state j at time t, which is subject to the following constraints.

Power balance constraints:

$$\boldsymbol{B}_j\cdot\boldsymbol{\theta}_j(t) = \boldsymbol{P}_j(t) - \boldsymbol{L}_j(t) \tag{30}$$

Generation limits:

$$\underline{P_{j_i}} \le P_{j_i}(t) \le \overline{P_{j_i}} \tag{31}$$

Load curtailment constraints:

$$0 \le LC_{j_i}(t) \le L_{j_i} \tag{32}$$

Line flow constraints:

$$|\frac{1}{x_j^{i\_i'}}(\theta_j^i - \theta_j^{i'})| \leq |\overline{F^{i\_i'}}| \tag{33}$$

$\boldsymbol{P}_j(t) = [P_{j_1}(t), \cdots, P_{j_N}(t)]^T$ and $\boldsymbol{L}_j(t) = [L_{j_1}(t), \cdots, L_{j_N}(t)]^T$ are the vectors of equivalent power generation and load for the state j at time t.

The loss of load probability (LOLP) and expected energy not supplied (EENS) is key indicators of evaluating the reliability of the power system, during operation period τ, the EENS and the LOLP are formulated in (34)–(37).

$$EENS_i(\tau) = \int_0^\tau \left(\sum_{j=1}^{K} p_{j_i}(t) \cdot LC_{j_i}(t) \right) dt \tag{34}$$

$$EENS(\tau) = \int_0^\tau \left(\sum_{j=1}^{K} p_{j_i}(t) \cdot \left(\sum_{i=1}^{N} LC_{j_i}(t) \right) \right) dt \tag{35}$$

$$LOLP_i(t) = \sum_{j=1}^{K} p_{j_i}(t) \cdot \mathbf{1}\left(LC_{j_i}(t) > 0\right) \tag{36}$$

$$LOLP(t) = \sum_{j=1}^{K} p_{j_i}(t) \cdot \mathbf{1}\left(\sum_{i=1}^{N} LC_{j_i}(t) > 0 \right) \tag{37}$$

where $\mathbf{1}(True) \equiv 1, \mathbf{1}(False) \equiv 0$.

4.2 Reliability Evaluation Procedure

The power system reliability assessment steps considering stochastic characteristics and cyber malfunctions are as follows:

Step 1: Power generation systems (containing wind farms and traditional generating units), cyber network systems, and VPPs are set in the initial state.

Step 2: The time-vary probabilities of the power generation system, the cyber network system, reserve capacity provided by VPPs and participation level of customers are calculated in different states.

Step 3: Formula (1)–(14) define the *Lz* transform of the actual available reserve capacity from VPPs, taking the cyber malfunctions factors into account.

Step 4: Formula (15)–(27) define the *Lz* transform of the whole generating units, taking the cyber malfunctions factors into account.

Step 5: Formula (28)–(33) define the Lz transform of the load curtailment at bus i.

Step 6: Formula (34)–(37) define the LOLP and the EENS as reliability assessment indicators.

5 System Studies and Cases

The modified IEEE RTS [31] is adopted for demonstrating the proposed models and techniques. A large wind farm consisting of 250 wind turbines with a rated power of 2 MW is added at bus 21 [9]. The Markov process model for the power output of a single wind turbine introduced in [32] is used for the studies. The MTTF and MTTR of a wind turbine are assumed to be 3650 and 55 h, respectively [9]. Five 575-MW coal thermal generating units modelled as a four-state Markov process [33] are installed in the system. The failure rate and repair rate including failures caused by cyber attacks of cyber systems in generation system are $8 * 10^{-4}$/h and 0.02/h [34], respectively. The failure rate and repair rate of cyber systems in demand side are 1/960 and 0.025 per hour, which are a little larger than those in the generation system. The system total demand L is 2850 MW. There is a proportion of interruptible loads can serve as VPPs to provide operating reserves in the study. The distribution of these VPPs at each bus is proportional to the base load in these buses. A four-state model for reserve capacity of VPPs is adopted where state transition rates are presented in Table 1. The multi-state model for participation level of customers presented in [4] is adopted. The system operation period is 100 h. The initial states for wind turbines, conventional generating units, and VPPs are the best states with the largest capacities.

5.1 Case 1: Different Reserve Capacities of VPPs

In Case 1, different proportions of DERs are aggregated in the form of VPPs and several scenarios shown in Table 2 are set according to different reserve capacities. At this time, there is no cyber network interruption or delay, and once the regulation signals are commanded, VPPs immediately provide operational reserve capacity.

Table 1 Transition rates of multi-state reserve capacity for VPPs

Transition rates (/h)	0	0.3 aL	0.6 aL	aL
0	–	0.039	0.013	0.008
0.3 aL	0.365	–	0.151	0.045
0.6 aL	0.122	0.220	–	0.192
aL	0.016	0.012	0.016	–

Table 2 The proportions of interruptible loads serving for different scenarios in case 1

Scenario	A	B	C	D	A'	B'
VPPs	0	0.1	0.2	0.4	0	0.1
Method	Proposed method				Simulation	

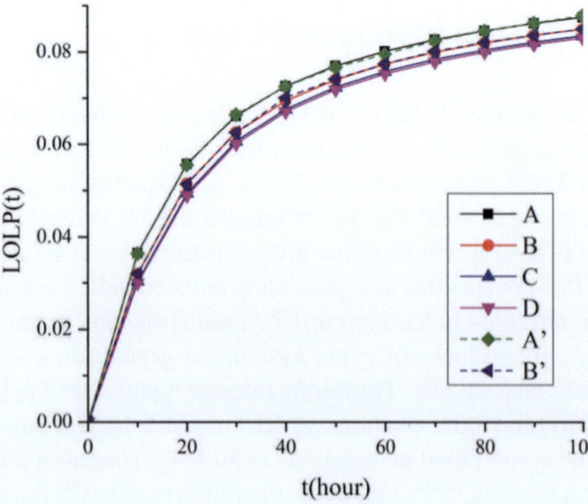

Fig. 5 Time-varying LOLP of bus 6 for case 1 for an operation duration of 100 h

The LOLP results for different scenarios represented at bus 6 are shown in Fig. 5, and these curves indicate that LOLP has time-varying characteristics rather than fixed values. LOLP is larger in Scenario A without VPPs. As the reserve capacity increases, the LOLP decreases slightly.

Furthermore, Monte Carlo simulations are used for comparative validation, where Scenarios A ' and B' are set based on the 10 0,000 sample size. The results of the proposed method have high computational accuracy, where the average error of the proposed method and simulation is relatively small at 0.37% (Results were obtained from a MatlabR2018b on a laptop with a 1.80 GHz processor).

The EENS results of the representative bus 6, the whole system and relative differences are presented in Table 3. Scenario A and D has a relative difference of 11.33%, and this data illustrates the increased reliability of systems with VPPs. Table 4 records the computing time of both methods, and the simulation method is much slower than the proposed method.

5.2 *Case 2: Malfunctions of Cyber Systems*

In Case 2, the proportion of load aggregated in VPPs of the total load is fixed as 20%, and the variable is the cyber malfunctions status. Set Scenario A as no cyber malfunctions, namely the system does not suffer cyber attacks or has no communication

Table 3 EENS and relative differences of the operation period for different scenarios in case 1

Scenario	EENS for bus (MWh)	System EENS (MWh)	Relative difference
A	293.3	6145.8	0
B	282.9	5929.0	−3.53%
C	274.4	5749.8	−6.44%
D	260.0	5449.3	−11.33%

Table 4 Computation time of different methods in case 1

Computation time (s)	Proposed method	Simulation
Scenario A	334.18	1537.9
Scenario B	335.67	1589.4
Scenario C	335.92	1591.1
Scenario D	335.99	1591.4

latency, Scenario B only considers cyber attacks, Scenario C considers cyber attacks and 2-h communication latency.

The LOLP results of the representative bus 6 and the ENNS results of whole system are presented in Fig. 6 and Table 5. The relative difference of Scenarios A and two other scenarios is over 60%. The system EENS for Scenarios B or Scenarios C is approximately 18.73 times that of Scenarios A. These data suggest that, cyber malfunctions will greatly weaken the system reliability. The comparison results in Scenarios B and Scenarios C are not obvious and will be discussed in Case 3.

Fig. 6 Time-varying LOLP of bus 6 for case 2 for an operation duration of 100 h

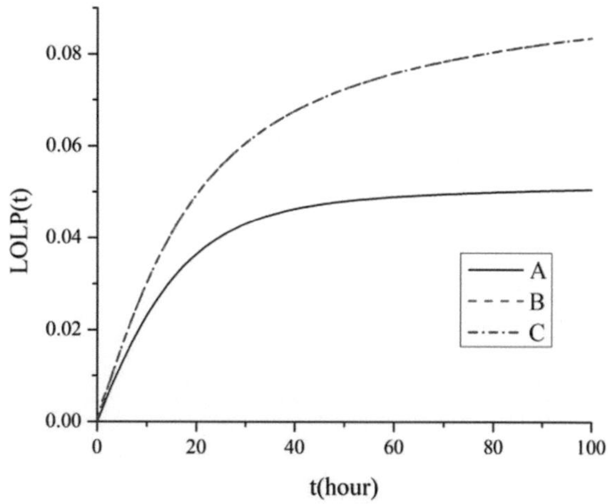

Table 5 System EENS for different scenarios in case 2

Scenario	A	B	C
System EENS (MWh)	307.01	5749.83	5750.80

5.3 *Case 3: Different Time for Providing Reserves from VPPs*

In Case 3, the proportion of load aggregated in VPPs of the total load is fixed as is 20%, and the variable is response time. On the basis of communication delay and response time, the VPPs commitment time is also taken into account. The above "time" factors are summarized as the time taken to provide reserve capacity in different scenarios, as shown in Table 6.

The LOLP results of the representative bus 6, the ENNS and relative differences results of operation period are presented in Fig. 7 and Table 7 respectively. VPPs are committed at once in Scenario A, with smallest system EENS. Once the the reserve capacity of VPPs is enabled, the LOLP begins to decrease. The longer the reserve capacity of VPPs is enabled, the greater the relative difference of EENS is.

Table 6 The time for providing reserve capacity of VPPs for different scenarios in case 3

Scenario	A	B	C	D	E	F
Time (h)	0	1	2	5	10	50

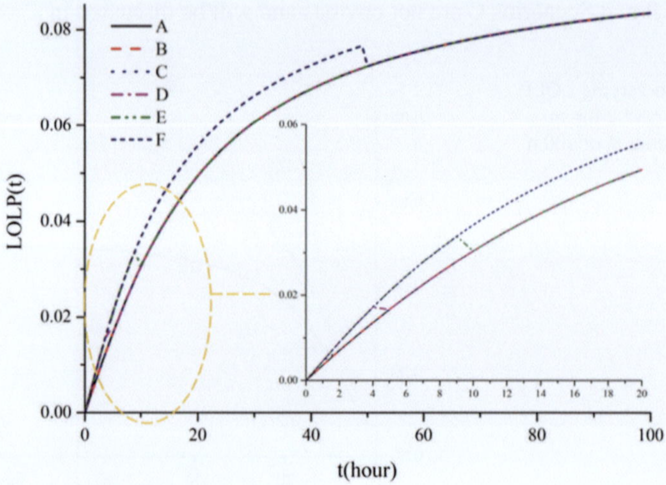

Fig. 7 Time-varying LOLP of bus 6 for case 3

Table 7 System EENS (MWh) and relative differences of the operation period in case 3

Scenario	A	B	C	D	E	F
EENS	5749.8	5750.1	5750.8	5755.5	5769.3	5929.7
Relative difference	0	0.004%	0.017%	0.098%	0.34%	3.13%

5.4 Case 4: Different Initial Conditions

In Case 4, the proportion of DERs which aggregates as VPPs is 20%. The VPPs commit immediately and the variable is initial conditions. The initial states for wind power generation and the reserve capacity of VPPs are listed at Table 8. In Scenario A, wind power and the reserve capacity are at their best states. In Scenario B, the reserve capacity is at the best state, while wind power is at the worst state. In Scenario C, the reserve capacity is at the the worst state, while wind power is at the best state. In Scenario D, both are at their worst states.

The system LOLP and EENS results are demonstrated in Fig. 8 and Table 9. Changes in the initial conditions had a large impact on the reliability. For example, the LOLP at $t = 10$ in Scenario D is 0.03995, which increases about 31.46% comparing with Scenario A. As the running time increases, The LOLP results for different scenarios progressively approach and converge with a certain value of the steady-state

Table 8 Initial states for different scenarios in case 4

Scenario	A	B	C	D
Wind power	Best state	Worst state	Best state	Worst state
VPPs	Best state	Best state	Worst state	Worst state

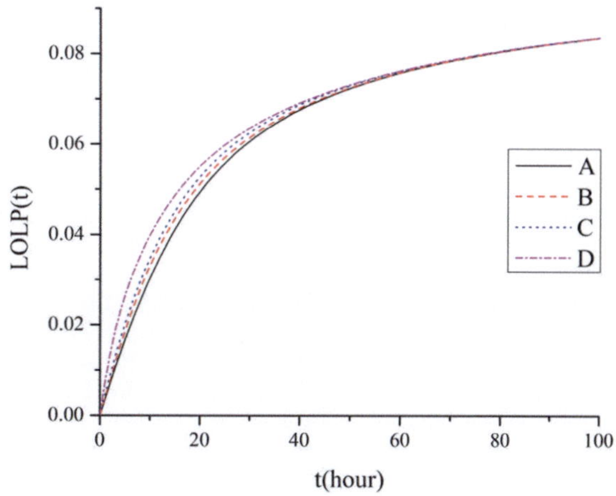

Fig. 8 System time-varying LOLP for case 4

Table 9 System EENS and relative differences of the operation period in case 4

Scenario	A	B	C	D
EENS (MWh)	5749.8	5809.3	5839.4	5917.6
Relative difference	0	1.04%	1.56%	2.92%

LOLP at 0.0924. The steady-state reliability assessment without studying the system initial conditions, so this is important for the time-varying reliability assessment.

References

1. A. Humayed, J. Lin, F. Li et al., Cyber-physical systems security—a survey. IEEE Internet Things J. **4**(6), 1802–1831 (2017)
2. H. Lei, C. Singh, Non-sequential Monte Carlo simulation for cyber-induced dependent failures in composite power system reliability evaluation. IEEE Trans. Power Syst. **32**(2), 1064–1072 (2017)
3. D. Xie, H. Hui, Y. Ding et al., Operating reserve capacity evaluation of aggregated heterogeneous TCLs with price signal. Appl. Energy **216**, 338–347 (2018)
4. H. Jia, Y. Ding, Y. Song et al., Operating reliability evaluation of power systems considering flexible reserve provider in demand side. IEEE Trans. Smart Grid **10**(3), 3452–3464 (2019)
5. J. Zhang, A.D. Domínguez-García, Evaluation of demand response resource aggregation system capacity under uncertainty. IEEE Trans. Smart Grid **9**(5), 4577–4586 (2018)
6. A.I. Negash, T.W. Haring et al., Allocating the cost of demand response compensation in wholesale energy markets. IEEE Trans. Power Syst. **30**(3), 1528–1535 (2015)
7. H. Hui, Y. Ding, Y. Song et al., Modeling and control of flexible loads for frequency regulation services considering compensation of communication latency and detection error. Appl. Energy **250**, 161–174 (2019)
8. G. Liang, S.R. Weller et al., The 2015 Ukraine Blackout: implications for false data injection attacks. IEEE Trans. Power Syst. **32**(4), 3317–3318 (2017)
9. Y. Ding, C. Singh, L. Goel et al., Short-term and medium-term reliability evaluation for power systems with high penetration of wind power. IEEE Trans. Sustain. Energy **5**(3), 896–906 (2014)
10. M. Moness, A.M. Moustafa, A survey of cyber-physical advances and challenges of wind energy conversion systems: prospects for internet of energy. IEEE Internet Things J. **3**(2), 134–145 (2016)
11. Y. Zhang, Y. Xiang et al., Power system reliability assessment incorporating cyber attacks against wind farm energy management systems. IEEE Trans. Smart Grid **8**(5), 2343–2357 (2017)
12. W. Liu, Q. Gong et al., Reliability modeling and evaluation of active cyber physical distribution system. IEEE Trans. Power Syst. **33**(6), 7096–7108 (2018)
13. C. Wang, T. Zhang, F. Luo et al., Impacts of cyber system on microgrid operational reliability. IEEE Trans. Smart Grid **10**(1), 105–115 (2019)
14. J. Guo, T. Zhao, W. Liu et al., Reliability modeling and assessment of isolated microgrid considering influences of frequency control. IEEE Access **7**, 50362–50371 (2019)
15. B. Hu et al., Decision-dependent uncertainty modeling in power system operational reliability evaluations. IEEE Trans. Power Syst. **36**(6), 5708–5721 (2021)
16. Y. Ding, P. Wang, L. et al., Long-term reserve expansion of power systems with high wind power penetration using universal generating function methods. IEEE Trans. Power Syst. **26**(2), 766–774 (2011)

17. Y. Ding, W. Cui et al., Multi-state operating reserve model of aggregate thermostatically-controlled-loads for power system short-term reliability evaluation. Appl. Energy **241**, 46–58 (2019)
18. G. Bao, Q. Li, J. Wang, et al., A comprehensive evaluation model of renewable energy benefits for power system planning. In *2019 IEEE 8th International Conference on Advanced Power System Automation and Protection (APAP)* (2019), pp. 986–990
19. M. Bao, Y. Ding, C. Singh et al., A multi-state model for reliability assessment of integrated gas and power systems utilizing universal generating function techniques. IEEE Trans. Smart Grid **10**(6), 6271–6283 (2019)
20. A. Lisnianski, Y. Ding, Using inverse Lz-transform for obtaining compact stochastic model of complex power station for short-term risk evaluation. Reliab. Eng. Syst. Saf. **145**, 19–27 (2016)
21. Y. Yan, Y. Ding, et al., Operating reliability analysis of peaking generating units considering start-up failures and degradation. In *2016 Second International Symposium on Stochastic Models in Reliability Engineering, Life Science and Operations Management (SMRLO)* (Beer Sheva, 2016), pp. 168–171
22. Y. Zhang, B. Wei, et al., Component importance indices of transmission systems based on the impact-increment based state enumeration method. In *2019 IEEE Innovative Smart Grid Technologies—Asia (ISGT Asia)* (2019), pp. 976–979
23. H. Wang, X. Liu et al., Probabilistic production simulation of a power system with wind power penetration based on improved UGF techniques. J. Mod. Power Syst. Clean Energy **1**(2), 186–194 (2013)
24. H. Chung, W. Li, C. Yuen et al., Local cyber-physical attack for masking line outage and topology attack in smart grid. IEEE Trans. Smart Grid **10**(4), 4577–4588 (2019)
25. R. Lai, X. Qiu et al., Robustness of asymmetric cyber-physical power systems against cyber attacks. IEEE Access **7**, 61342–61352 (2019)
26. D. Liu, C.K. Tse, Cascading failure of cyber-coupled power systems considering interactions between attack and defense. IEEE Trans. Circuits Syst. I Regul. Pap. **66**(11), 4323–4336 (2019)
27. M.A. Shahab, B. Mozafari et al., Stochastic consensus-based control of μGs with communication delays and noises. IEEE Trans. Power Syst. **34**(5), 3573–3581 (2019)
28. T. Ramachandran, M.H. Nazari et al., Overcoming communication delays in distributed frequency regulation. IEEE Trans. Power Syst. **31**(4), 2965–2973 (2016)
29. K.R. Davis, C.M. Davis et al., A cyber-physical modeling and assessment framework for power grid infrastructures. IEEE Trans. Smart Grid **6**(5), 2464–2475 (2015)
30. H. Jia, C. Shao, D. Liu, C. Singh, Y. Ding, Y. Li, Operating reliability evaluation of power systems with demand-side resources considering cyber malfunctions. IEEE Access **8**, 87354–87366 (2020)
31. Reliability Test System Task Force of the Application of Probability Methods Subcommittee, "IEEE reliability test system. IEEE Trans. Power Apparat. Syst. **98**(6), 2047–2054 (1979)
32. A.S. Dobakhshari, M. Fotuhi-Firuzabad, A reliability model of large wind farms for power system adequacy studies. IEEE Trans. Energy Convers. **24**(3), 792–801 (2009)
33. A. Lisnianski, D. Elmakias et al., A multi-state Markov model for a short-term reliability analysis of a power generating unit. Reliab. Eng. Syst. Saf. **98**(1), 1–6 (2012)
34. M. Panteli, D.S. Kirschen, Assessing the effect of failures in the information and communication infrastructure on power system reliability. In *2011 IEEE/PES Power Systems Conference and Exposition* (Phoenix, AZ, 2011), pp. 1–7

FSC
www.fsc.org
MIX
Papier aus verantwortungsvollen Quellen
Paper from responsible sources
FSC® C105338

If you have any concerns about our products,
you can contact us on
ProductSafety@springernature.com

In case Publisher is established outside the EU,
the EU authorized representative is:
Springer Nature Customer Service Center GmbH
Europaplatz 3, 69115 Heidelberg, Germany

Printed by Libri Plureos GmbH
in Hamburg, Germany